# Polymer Nanocomposite Foams

# Polymer Nanocomposite Foams

Edited by
## Vikas Mittal

CRC Press
Taylor & Francis Group
Boca Raton London New York

CRC Press is an imprint of the
Taylor & Francis Group, an **informa** business

CRC Press
Taylor & Francis Group
6000 Broken Sound Parkway NW, Suite 300
Boca Raton, FL 33487-2742

First issued in paperback 2017

© 2014 by Taylor & Francis Group, LLC
CRC Press is an imprint of Taylor & Francis Group, an Informa business

No claim to original U.S. Government works

ISBN-13: 978-1-4665-5812-0 (hbk)
ISBN-13: 978-1-138-07499-6 (pbk)

**Library of Congress Cataloging-in-Publication Data**

Polymer nanocomposite foams / editor, Vikas Mittal.
     pages cm
   Includes bibliographical references and index.
   ISBN 978-1-4665-5812-0 (hbk. : alk paper)
    1. Plastic foams. 2. Plastic foams. 3. Inorganic polymers. 4. Nanocomposites (Materials) I. Mittal, Vikas.

TP1183.F6P645 2013
668.4'93--dc23                                                     2013030658

**Visit the Taylor & Francis Web site at
http://www.taylorandfrancis.com**

**and the CRC Press Web site at
http://www.crcpress.com**

# Contents

# Preface

This book focuses on the subject of foams generated with polymer nanocomposite materials. Polymer nanocomposites have been developed continuously for the last two decades. These advancements have led to their application in many fields such as automotive, packaging, insulation, and so forth. Foams are one product, which is common to many application fields and also has high commercial value. Use of nanocomposites in the formation of foams enhances the property profiles such as porosity control, strength, stiffness, and so on, significantly, which enables the application of such materials in conventional areas ranging to more advanced ones.

The generation of nanostructured foams is affected by a large number of factors such as the nature of the polymer, the methods used to achieve the cellular structure, the interaction of the polymer with the filler surface, the dispersion state of the filler, and so forth. A small change in the process variables completely affects the structure and properties of the resulting foams, thus a thorough understanding of the various factors affecting the foams' structure–property correlations is needed. The book aims to compile the advancements in the various aspects of nanocomposite foams with the objective of providing background information to readers new to this field as well as to serve as a reference text for researchers in this area.

In Chapter 1, different synthesis and processing techniques used to prepare poly(methyl methacrylate) (PMMA) nanocomposite foams are reviewed. The effects of nanoparticles on foam morphology and properties are discussed.

In Chapter 2, the strategies of toughening polymer foams, particularly rigid polymer syntactic foams, are discussed. A general introduction of toughness and the toughening mechanism in brittle polymer systems is provided followed by a description of successful and effective toughening strategies for thermoset/hollow sphere syntactic foams using microfibers and nanoparticles. In Chapter 3, the effect of nanoclay addition on the foaming behavior of polypropylene is summarized. In general, the presence of well-dispersed nanoclay-enhanced cell nucleation and suppressed cell coalescence, result in a significant increase in cell density by two or three orders of magnitude and foam expansion. Chapter 4 describes various routes, such as fiber and nanoclay incorporation, to starch foams to improve the performance of these materials. Chapter 5 reviews the recent progress in achieving lightweight polymer nanocomposite foams with high performance without sacrificing mechanical properties. Chapter 6 focuses on the development and main properties of hybrid polyurethane nanocomposite foams, flexible as well as rigid, focusing on the influence of processing and incorporation of various types of nanometric-sized fillers in the structure and mechanical properties, transport properties, and other significant properties of the resulting foams. In Chapter 7, nanocomposite morphologies, types of polymer-clay nanocomposite production, and modifications in polymer and montmorillonite structures, which allow them to be used in nanocomposite preparation, are explained. In addition, the concepts involving foam production and its morphology are presented. Chapter 8 discusses recent advances in the field of carbon nanotube/polymer nanocomposite aerogels and related materials. There is an emphasis on the relationship between the preparation method and the most characteristic properties of these materials such as density, surface area, electrical conductivity, mechanic strength, and so on. The book concludes with Chapter 9, which presents a review of the nanocomposite foams generated from high-performance thermoplastics.

**Vikas Mittal**
*Abu Dhabi*

# Editor

**Vikas Mittal, Ph.D.,** is an assistant professor in the Chemical Engineering Department of The Petroleum Institute, Abu Dhabi. He obtained his Ph.D. in 2006 in polymer and materials engineering from the Swiss Federal Institute of Technology in Zurich, Switzerland. Dr. Mittal then worked as a materials scientist in the Active and Intelligent Coatings section of Sun Chemical in London, and as a polymer engineer at BASF Polymer Research in Ludwigshafen, Germany. His research interests include polymer nanocomposites, novel filler surface modifications, thermal stability enhancements, and polymer latexes with functionalized surfaces, among others. Dr. Mittal has authored more than 50 scientific publications, book chapters, and patents on these subjects. (E-mail: vik.mittal@gmail.com)

# Contributors

**Priscila Anadão**
Metallurgical and Materials
  Engineering Department
School of Engineering
University of São Paulo
Cidade Universitária
São Paulo, Brazil

**Marcelo Antunes**
Centre Català del Plàstic
Departament de Ciència dels Materials i
  Enginyeria Metal·lúrgica
Universitat Politècnica de Catalunya
BarcelonaTech (UPC)
Terrassa, Barcelona, Spain

**Zhenhua Chen**
High-Performance Materials Institute
Florida State University
Department of Industrial and
  Manufacturing Engineering
Florida A&M University
Florida State University College of
  Engineering
Tallahassee, Florida

**Jannick Duchet-Rumeau**
Université de Lyon
Lyon, France
CNRS
Ingénierie des Matériaux Polymères
INSA Lyon
Villeurbanne, France

**Maria Victoria E. Grossmann**
Department of Food Science and
  Technology
State University of Londrina
Paraná, Brazil

**Xiao Hu**
School of Materials Science and
  Engineering
Nanyang Technological University
Singapore

**Salvatore Iannace**
Institute for Composite and Biomedical
  Materials
National Research Council of Italy
Portici, Italy

**Yan Li**
High-Performance Materials Institute
Florida State University
Department of Industrial and
  Manufacturing Engineering
Florida A&M University
Florida State University College of
  Engineering
Tallahassee, Florida

**Ming Liu**
School of Materials Science and
  Engineering
Nanyang Technological University
Singapore

**Sebastien Livi**
Université de Lyon
Lyon, France
CNRS
Ingénierie des Matériaux Polymères
INSA Lyon
Villeurbanne, France

**Suzana Mali**
Department of Biochemistry and
  Biotechnology
State University of Londrina
Paraná, Brazil

**Petar Dimitrov Petrov**
Bulgarian Academy of Sciences
Institute of Polymers
Sofia, Bulgaria

**Luigi Sorrentino**
Institute for Composite and Biomedical
    Materials
National Research Council of Italy
Portici, Italy

**Kun Wang**
Ningbo Key Lab of Polymer Materials
Ningbo Institute of Material Technology
    and Engineering
Chinese Academy of Sciences, Ningbo
Zhejiang, China

**Erwin M. Wouterson**
School of Mechanical and Aeronautical
    Engineering
Singapore Polytechnic
Singapore

**Fabio Yamashita**
Department of Food Science and
    Technology
State University of Londrina
Paraná, Brazil

**Changchun Zeng**
High-Performance Materials Institute
Florida State University
Department of Industrial and
    Manufacturing Engineering
Florida A&M University
Florida State University College of
    Engineering
Tallahassee, Florida

**Wentao Zhai**
Ningbo Key Lab of Polymer Materials
Ningbo Institute of Material Technology
    and Engineering
Chinese Academy of Sciences, Ningbo
Zhejiang, China

**Liying Zhang**
School of Materials Science and
    Engineering
Nanyang Technological University
Singapore

# 1 Poly(Methyl Methacrylate) (PMMA) Nanocomposite Foams

*Yan Li, Zhenhua Chen, and Changchun Zeng*

## CONTENTS

## 1.1 INTRODUCTION

PMMA is an important polymer for mechanical and optical applications due to its feasibility, good tensile strength and hardness, high rigidity, high transparency in the visible wavelength range, high surface resistivity, good insulation properties, and thermal stability. In the last two decades, PMMA nanocomposites incorporating nanoscale particles have attracted increasing attention from both academia and industry because of the high potential for new and/or improved properties enabled and/or enhanced by these nanoparticles (Burda et al., 2005).

1

Separately, intense research activities in PMMA foams have generated a class of lightweight and cost friendly materials. Their porous structure makes them ideal for a variety of applications ranging from packaging, insulation, cushions, and adsorbents to scaffolds for tissue engineering (Darder et al., 2011; Zeng et al., 2010). Moreover, the rapid development of applying supercritical carbon dioxide foaming technology promises an environmentally friendly process, compared with the traditional chloro-fluorocarbon (CFC) foaming methods. However, the applications of these foams are limited by their inferior mechanical strength, poor surface quality, and low thermal and dimensional stability (Lee et al., 2005).

In recent years, novel PMMA nanocomposite foams have been investigated as an emerging and interdisciplinary topic at the boundary between materials science, process, and nanotechnology. The combination of functional nanoparticles and porous structure enable their versatile use as new materials that are lightweight and have a high strength-to-weight ratio and well-defined functions or are multifunc-tional (Lee et al., 2005; Siripurapu et al., 2005; Sun, Sur, and Mark, 2002; Zeng et al., 2003).

In this chapter, we summarize the highlights of the major developments in this area during the last decade. First, the different synthesis and processing techniques used to prepare PMMA nanocomposites are briefly reviewed. This is followed by a brief review of foaming processing methods. The effects of nanoparticles on the foam morphology and properties are then discussed in great detail. Finally, the pro-cessing and application of PMMA nanocomposite foams are addressed.

## 1.2   SYNTHESIS OF PMMA NANOCOMPOSITES

The enormous interest in using nanoparticles in polymer matrices is due to the exceptional potential to enhance a wide range of properties, such as electrical conductivity, thermal stability, mechanical enhancement and strength, and barrier performance. Typically, nanoparticles can be classified as three different types (Ashby, Ferreira, and Schodek, 2009): (1) zero-dimensional (0D), (2) one-dimensional (1D), and (3) two-dimensional (2D), as shown in Figure 1.1. 0D nanoparticles are materials where all dimensions are in nanometer scale, for example, spherical silica particles (Chen et al., 2004; Goren et al., 2010; Yang et al., 2004). A variant of this type of particle are the highly porous particles. While the dimension of the particle may be in the order of microns, the pore size is in the order of nanometers (Luo, 1998). 1D nanoparticles have two dimensions in the nanometer regime (< 100 nm) and the typical particles include nanowires, nanorods, nanofibers, and nanotubes (Chen, L., et al., 2010, 2011, 2012; Chen, Z., et al., 2011; Gorga et al., 2004; Zeng et al., 2010, 2013). The third type of nanoparticles is 2D nano-materials, which only have one dimension in the nanometer scale which are platelet like. Nanoclay (Jo, Fu, and Naguib, 2006; Zeng et al., 2003) and graphene (Ramanathan et al., 2007; Zhang et al., 2011) are good examples of this type of nanoparticles. All these nanoparticles have been used in PMMA nanocomposite foams.

The ubiquitous challenge in polymer nanocomposite preparation, that is, estab-lishing a good nanoparticle dispersion in the host polymer matrix, is also the main issue for PMMA nanocomposite fabrication (Bauhofer and Kovacs, 2009; Chatterjee, 2010; Moniruzzaman and Winey, 2006). Good dispersion of the fillers is

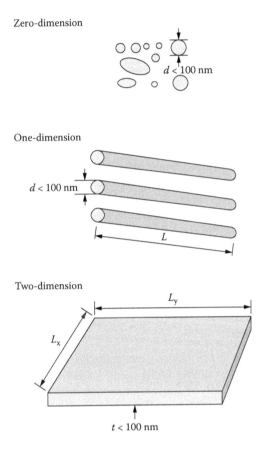

Zero-dimension

$d < 100$ nm

One-dimension

$d < 100$ nm

$L$

Two-dimension

$L_y$

$L_x$

$t < 100$ nm

**FIGURE 1.1** Schematic of different nanoparticles.

important to realize the exceptional properties of the fillers in the nanocomposites. However, this is a difficult task because of the high specific surface area and strong intermolecular forces associated with these nanoparticles. Moreover, since the predominant nanoparticles are inorganic and the surfaces are usually hydrophilic, they need to be modified/functionalized for improving interaction and compatibility with the typically hydrophobic polymers. Of equal importance are the nanocomposite preparation methods, which oftentimes need to be optimized in conjunction with nanoparticle surface functionalization to achieve good particle dispersion. In this section, we briefly discuss the most common methods for nanocomposite processing and their applications in PMMA nanocomposite preparation.

### 1.2.1 Solution Blending

In solution blending, a solvent or solvent mixture is employed to disperse the nanoparticles and dissolve PMMA (Moniruzzaman and Winey, 2006; Zeng et al., 2010). The common problem with most processing methods is proper dispersion of the nanoparticles in solvents. Choosing a good solvent system is important to the separation of the nanoparticles due to the weak van der Waals interactions where the polymer chains are able to coalesce with the nanoparticles (Lee et al., 2005). Nanoparticles typically agglomerate or cluster together during and after processing, causing nonuniform dispersion within the polymer matrix. To address this, typically solution blending is done by sonication, which uses sound waves to separate nanoparticle clusters in liquid solvents. Once the sonication is complete, the PMMA nanocomposites can be prepared using two methods: solvent casting (SC) and antisolvent precipitation (ASP). In the SC process, the PMMA-nanoparticle-solvent mixture is casted and nanocomposites are obtained after solvent drying. In the ASP process, an antisolvent is added to the mixture and the polymer nanocomposite precipitates. It is then collected and dried.

Du, Fischer, and Winey (2003) and Zeng et al. (2010) used both methods in attempts to produce the PMMA/carbon nanotube (CNT) nanocomposites. They found ASP resulted in better CNTs dispersion. Unlike the solvent casting process where nanoparticles may agglomerate during solvent evaporation, in ASP, the rapid precipitation of PMMA-CNTs very effectively lock down the well-dispersed structure.

### 1.2.2 MELT BLENDING

Instead of using solvent as the medium, nanoparticles can be mixed directly with a molten PMMA either statically or under shear. Unlike the solution blending, melt blending does not require solvents, and is compatible with industrial polymer extrusion and blending processes. Thus, it offers an economically attractive route in fabricating polymer nanocomposites (Lee et al., 2005). However, very careful attention needs to be paid to finely tune the nanoparticles' surface chemistry to increase the compatibility with the polymer matrix. In addition, processing conditions have profound effects on the structure evolution of polymer nanocomposites (Wang et al., 2001). Control of the shear force is essential in order not to damage the nanoparticles and degrade the nanocomposite properties (Lee et al., 2005).

Intercalated PMMA/clay nanocomposites were prepared by melt mixing (Kumar, Jog, and Natarajan, 2003; Zeng et al., 2003) using organically modified nanoclays. Upon PMMA intercalation, interlayer spacing was expanded as confirmed by X-ray diffraction. Wang and Guo (2010) reported the synthesis of PMMA/clay nanocomposites with styrene-maleic anhydride copolymers (SMA). SMA and the required amount of clay were dry mixed and then fed into the molten PMMA, and melt blended. As evidenced by X-ray diffraction (XRD), the organoclay was well intercalated in the PMMA matrix. The transmission electron microscopy (TEM) studies also showed that the nanoclay was intercalated and randomly dispersed in the PMMA matrix.

### 1.2.3 IN SITU POLYMERIZATION

Another technique that has been used to make PMMA nanocomposites is *in situ* polymerization since the 1960s (Blumstein, 1965; Huang and Brittain, 2001; Lee and Jang, 1996). It is a method involving dispersing nanoparticles in a monomer followed by polymerization of the solution (Zeng and Lee, 2001). In comparison to solution blending, *in situ* polymerization uses little or no solvent. The low viscosity of monomer (compared to melt viscosity) is beneficial for mixing and better dispersion of fillers, making *in situ* polymerization an attractive route for nanocomposite synthesis. On the other hand, the process is more complicated and more difficult to implement.

Zeng and Lee (2001) prepared PMMA/clay nanocomposites via *in-situ* bulk polymerization. The compatibility of the initiator and monomer with the clay surface was found to profoundly affect the clay dispersion. Furthermore, by using a nanoclay (MHABS) that was modified by a surfactant containing a polymerizable group (the chemical structure is shown in the top right of Figure 1.2), exfoliated PMMA/clay nanocomposites with excellent clay dispersion were synthesized.

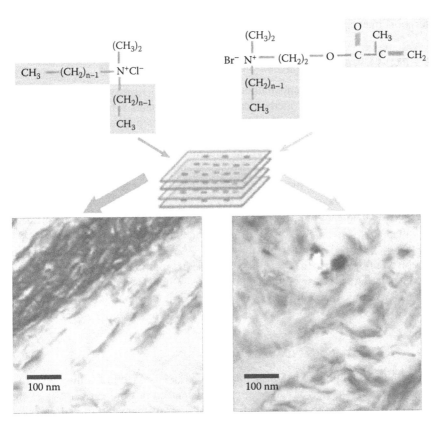

**FIGURE 1.2** Intercalated (PMMA/20A) and exfoliated (PMMA/MHABS) nanocomposites. Shown on top are the surfactants to modify the nanoclay (middle). Note the acrylate double present in the surfactant on the right. Shown on bottom are TEM micrographs showing the nanoclay dispersion. (Reprinted with permission from Zeng C. et al., *Advanced Materials* 2003, 15, 1743–1747. Copyright 2003, John Wiley & Sons; Reprinted with permission from Lee L. et al., *Composites Science and Technology* 2005, 65, 2344–2363. Copyright 2005, Elsevier.)

Wang et al. (2002) compared various *in situ* polymerization methods for the preparation of PMMA/clay nanocomposites. It was found that the particular preparative technique that is used has a large effect on the type of nanocomposites (in terms of nanoclay dispersion) that may be obtained. Solution polymerization of MMA only yields intercalated nanocomposites regardless of the presence of polymerizable double bond in the intergallery region. On the other hand, emulsion, suspension, and bulk polymerization can yield either exfoliated (with intergallery double bond) or intercalated (without double bond present) nanocomposites.

Yeh et al. (2009) prepared PMMA/organoclay nanocomposite systems by *in situ* polymerization using benzoyl peroxide (BPO) as the initiator. It was found that when mixed, an intercalated–exfoliated structure of nanocomposite material was formed. The molecular weights of extracted PMMA were found to be significantly lower than that of neat PMMA, indicating polymerization is structurally confined in the intragallery region of the clay, and the nature of clay–oligomer interactions, such as adsorption, may play a role during polymerization.

## 1.3   PMMA NANOCOMPOSITE FOAM PREPARATION

The synthesized PMMA nanocomposites can be used to produce PMMA nanocomposite foams. The main method used to produce foams is the direct utilization of foaming agents. Two types of foaming agents are often used: chemical blowing agents (CBAs) or physical blowing agents (PBAs). Almost exclusively, studies to-date focus on foaming using a physical blowing agent. In particular, supercritical carbon dioxide ($scCO_2$) as a physical blowing agent has attracted wide attention due to its marked advantages, such as low cost, environmental benignancy, and easily accessible supercritical conditions ($T_c = 31°C$, $P_c = 7.38$ MPa), as well as the tunability of physicochemical properties (such as density and mobility) by varying pressure and temperature (Cooper, 2000; Johnston and Shah, 2004).

Typically, physical foaming is a three-step process: (1) mixing: a blowing gas is dissolved in the polymer to form a homogeneous solution; (2) bubble nucleation: subsequent pressure release or temperature increase induces phase separation due to the thermodynamic instability, and gas starts to form nuclei; and (3) bubble growth and stabilization.

### 1.3.1   NONCONTINUOUS FOAMING

Noncontinuous foaming, or batch foaming, is commonly used in foaming research. In batch foaming, the polymer nanocomposite is placed in a pressurized vessel and saturated with the foaming agent under predetermined temperature and pressure. If the temperature is higher than the glass transition temperature, $T_g$, the release of pressure would result in supersaturation and cell nucleation and growth. Cell structure is usually fixed by cooling the materials to a temperature below the $T_g$. This is commonly referred to *pressure quench technique*. On the other hand, when the saturation temperature is lower than $T_g$, the cell is unable to nucleate and grow after the release of pressure even when the gas is in the supersaturation state because of the glassy nature (high rigidity) of the matrix. Foaming may occur when temperature

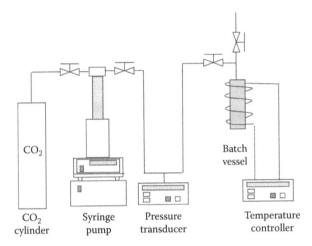

**FIGURE 1.3** Schematic of a typical batch foaming setup. (Reprinted with permission from Zeng C. et al., *Polymer* 2010, 51, 655–664. Copyright 2010, Elsevier.)

is raised above $T_g$. This is referred to as *temperature jump technique.* Both are routinely used in batch foaming studies. Cell structure is again fixed by cooling. Batch foaming is usually carried out at temperatures far below the polymer flowing temperature. The saturation time is usually very long (from hours to days depending on the gas diffusivity). This greatly limits the productivity. Figure 1.3 shows a typical high pressing foaming system.

### 1.3.2 Continuous Foaming

Continuous extrusion foaming is the most commonly used technology in the foam industry. Continuous foaming is used through the extrusion method. Both single- and twin-screw extruders can be used for plastic foaming. A schematic of a typical extrusion foaming system is shown in Figure 1.4 (Han et al., 2003). Multiple temperature zones and pressure sensors may be implemented. Extrusion foaming is performed by injecting a foaming gas (typically by a syringe pump for precise metering) into an extrusion barrel, combined with the polymer nanocomposite. When the homogenous polymer/gas mixture passes through a die, a rapid pressure drop induces phase separation and cell nucleation. Pressure drop rate is particularly important in controlling cell nucleation. A shaping die can be used to control the product shape and foam expansion. The foamed materials continue to expand until the extrudate temperature is lower than $T_g$ and the foam product is vitrified.

### 1.3.3 Retrograde Foaming

Most polymer-gas systems have a single glass transition temperature at a given gas pressure or gas concentration, which often decreases linearly with gas pressure or gas concentration in polymers. PMMA-$CO_2$ is one of the few polymer systems that

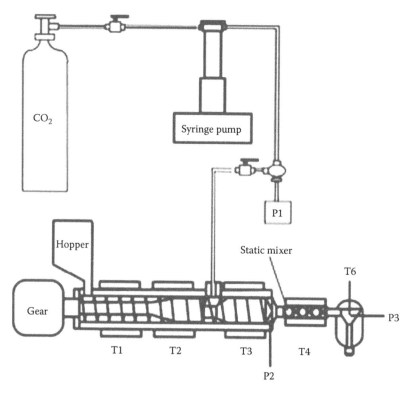

**FIGURE 1.4** Schematic of a typical continuous extrusion foaming setup. (Reprinted with permission from Han X. et al., *Polymer Engineering and Science* 2003, 43, 1261–1275. Copyright 2003, John Wiley & Sons.)

exhibit a unique phenomenon: retrograde vitrification. Due to the intricate behavior and interplay between the solubility and the resultant plasticization and reduction of glass transition temperature ($T_g$) by the dissolved carbon dioxide, these polymer $CO_2$ systems possess two $T_g$s (Condo, Paul, and Johnston, 1994; Handa and Zhang, 2000; Handa, Zhang, and Wong, 2001). Shown in Figure 1.5a is the glass transition temperature as a function of $CO_2$ pressure for a PMMA-$CO_2$ system (Handa and Zhang, 2000), where two $T_g$s exist over a wide pressure range. Upon being cooled below the low $T_g$, the systems change from glassy state to rubbery state. In the retrograde phase, the solubility in PMMA is exceptionally high. Furthermore, the rubbery state ensures possible foamability. Both will be beneficial for producing foams with exceptionally high cell density and small cell size. Indeed, PMMA foams with exceptionally high cell density were prepared from the retrograde phase (Handa and Zhang, 2000; Handa, Zhang, and Wong, 2001). Shown in Figure 1.5b is a PMMA foam prepared by retrograde foaming by our group (Chen, Z., 2011). The foam exhibits high cell density ($10^{11}$ cells/cm$^3$) and small cell size (average 1~2 µm). The PMMA foam also exhibits a fairly uniform cell size. Even small size and higher cell density were reported in the literature (Handa and Zhang, 2000; Handam, Zhang, and Wong, 2001).

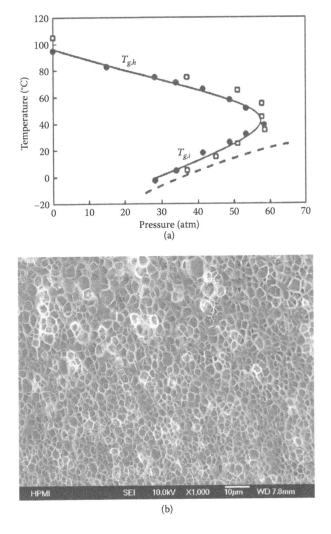

**FIGURE 1.5** PMMA foaming from the retrograde region. (a) Glass transition temperature as a function of $CO_2$ pressure; dashed line is the vapor–liquid phase boundary. (Reprinted with permission from Handa Y. P. and Zhang Z., *Journal of Polymer Science: Part B: Polymer Physics* 2000, 38, 716–725. Copyright 2000, John Wiley & Sons.) (b) SEM micrograph of a PMMA foam prepared by foaming from retrograde phase. (Reprinted with permission from Chen Z. et al., *SPE ANTEC* 2011, 69, 2678–2682. Copyright 2011, Society of Plastics Engineers.)

Retrograde foaming of PMMA nanocomposites have been studied. Zeng et al. (2003) prepared clay nanocomposite foam by retrograde foaming of PMMA-5%. MHABS nanocomposite and submicron cellular foams were prepared (Figure 1.6).

Chen et al. (2011) prepared PMMA-CNT nanocomposite foams by retrograde foaming and identified two additional complications that might occur. First, as will be discussed in great detail in Section 1.4.1.2, the exceptionally high nucleation rates

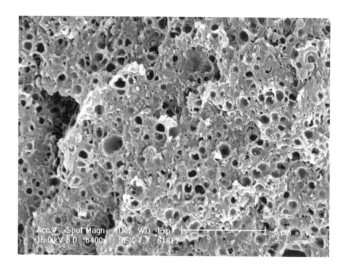

**FIGURE 1.6** SEM of a PMMA/5% MHABS nanocomposite foam by retrograde foaming. The average cell size is around 0.3 μm and the cell density is around is $1.86 \times 10^{12}$ cells/cm$^3$. (Reprinted with permission from Zeng C. et al., *Advanced Materials* 2003, 15, 1743–1747. Copyright 2003, John Wiley & Sons.)

impose significantly more stringent requirements on the nanoparticle dispersion in order to obtain a uniform cell size distribution; second, while a strong polymer–CNT interaction is beneficial for nanoparticle dispersion, when coupled with the relatively low foaming temperature used, significantly increases matrix rigidity (which by itself is already very high as a relatively low temperature is used in retrograde foaming) as the result of well-dispersed nanoparticles, and foaming may be prohibited. Such phenomenon was observed in the retrograde foaming of PMMA-2wt% CNT nanocomposite foams (Figure 1.7).

## 1.4 MORPHOLOGY AND PROPERTIES

Recently, foaming of polymer nanocomposites has emerged as a novel means to expand the accessible range of foam morphology, and produce novel multifunctional materials with enhanced properties (Ibeh and Bubacz, 2008; Lee et al., 2005). The impact of nanoparticles on the polymer foams are mainly twofold: (1) alteration of morphology resulting from the introduction of nanoparticles; and (2) change of properties as a combined effect of morphological change and properties enabled/ enhanced by the nanoparticles.

The properties of the polymer nanocomposites are dictated by the types of nanoparticle used and the foam morphology. The foam morphology, in turn, is largely determined by the nanocomposite synthesis (nanoparticle dispersion) and foaming conditions. Due to the complicated nature of the interactions between nanoparticles, bubbles, and matrix, the influence of nanoparticles on the properties of nanocomposite foams is still not fully understood (Bauhofer and Kovacs, 2009; Chen, Ozisik, and Schadler, 2010; Moniruzzaman and Winey, 2006).

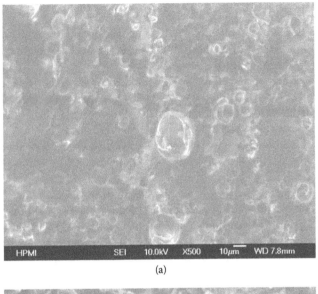

(a)

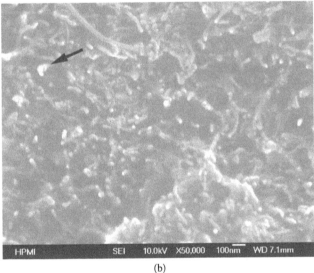

(b)

**FIGURE 1.7**   SEM of a PMMA-2wt% CNT nanocomposite foamed from the retrograde phase (a) low and (b) high magnifications. No cellular morphology was observed and foaming was prohibited. Note that cellular morphology was obtained at the same foaming conditions for neat PMMA (Figure 1.5b) and PMMA-0.5wt% CNT nanocomposite. The CNTs interact strongly with PMMA to form a core-sheath structure where PMMA wraps around the nanotube. The arrow in (b) indicates one such structure. (Reprinted with permission from Chen Z. et al., *SPE ANTEC* 2011, 69, 2678–2682. Copyright 2011, Society of Plastics Engineers.)

## 1.4.1 Morphology

Nanoparticles are now commonly used to foam cell morphology manipulation because they significantly affect both cell nucleation and cell growth, the two most important processes in foaming.

Nanoparticles are highly effective bubble nucleating agents, leading to foams with higher cell density and smaller cell size. This has been observed in numerous foams utilizing different types of nanoparticles: this results in the reduced cell size in the foams because the available gas for bubble growth is lowered as a greater number of nucleated bubbles grow simultaneously. Moreover, the nanoparticles can significantly increase the melt viscosity, which hinders cell growth and leads to a reduced cell size.

The high nucleation efficiency of nanoparticles has been shown to be particularly advantageous for manufacturing microcellular foam (cell size <10 μm, cell density >$10^9$ cells/cc³) (Martini-Vvedensky and Waldman et al., 1982). The nucleation efficiency of the nanoparticles is dependent on the particle geometry, aspect ratio, dispersion, concentration, and particle surface treatment. These are discussed in detail in this section. The resulting changes in foam structure (bubble density, bubble size, and size distribution) and matrix properties have profound influence on the foam mechanical properties.

### 1.4.1.1 Effect of Nanoparticle Geometry and Concentration

Compared to conventional microsized filler particles used in the foaming processes, nanoparticles offer unique advantages for enhanced nucleation. The extremely fine dimensions and large surface area of nanoparticles provide much more intimate contact between the fillers, polymer matrix, and gas. Furthermore, a significantly higher effective particle concentration can be achieved at a low nominal particle concentration. Both could lead to improved nucleation efficiency.

While the efficiency of nanoparticles for enhancing nucleation has been widely reported and superior to micron-sized particles, the effects of particle size and geometry in general (shape, aspect ratio, and surface curvature) require further elucidation.

Fletcher (1958) ascertained the effects of particle geometry on the nucleation efficiency and this is briefly summarized below.

Based on the classical nucleation theory (Abraham 1974; Laaksonen, Talanquer, and Oxtoby, 1995), the heterogeneous nucleation rate is expressed as

$$N_{het} = v_{het} C_{het} \exp(-\Delta G^*_{het}/kT) \tag{1.1}$$

where $C_{het}$ is the concentration of heterogeneous nucleation sites, $k$ is the Boltzmann constant, $T$ is temperature, $v_{het}$ is the frequency factor of gas molecules merging with the nucleus, and $\Delta G^*_{het}$ is the critical Gibbs free energy to form a critical embryo on the nucleating sites, that is,

$$\Delta G^*_{het} = \frac{\Delta G^*_{hom}}{2} f(m, w) \tag{1.2}$$

$$\Delta G^*_{\text{hom}} = \frac{16\pi\gamma_{lv}^3}{3\Delta P^2} \tag{1.3}$$

$$\gamma_{lv} = \gamma_{lv0}\left[1 - \frac{T}{T_c}\right]^{11/9} \tag{1.4}$$

$$f(m,w) = 1 + \left(\frac{1-mw}{g}\right)^3 + w^3\left[2 - 3\left(\frac{w-m}{g}\right) + \left(\frac{w-m}{g}\right)^3\right] + 3mw^2\left(\frac{w-m}{g} - 1\right) \tag{1.5}$$

$$m = \cos\theta \tag{1.6}$$

$$w = R/r^* \tag{1.7}$$

$$r^* = \frac{2\gamma_{lv}}{\Delta P} \tag{1.8}$$

$$g = (1 + w^2 - 2mw)^{\frac{1}{2}} \tag{1.9}$$

where $\Delta G^*_{\text{hom}}$ is the homogeneous Gibbs free energy, which is a function of polymer-gas surface tension, $\gamma_{lv}$, and the pressure difference ($\Delta P$) between that inside the critical nuclei and that around the surrounding liquid. Assuming that the polymer is fully saturated by $CO_2$ and the partial molar volume of $CO_2$ in the polymer is zero, can be considered to be the difference between the saturation pressure and atmospheric pressure. $f$ is the critical energy reduction factor due to the inclusion of nucleants, which is a function of the polymer-gas-particle contact angle $\theta$ and the relative curvature of the nucleant surface to the critical radius of the nucleated phase (Equations 1.6 through 1.10). $r^*$ is the critical radius.

Using the above set of equations, the critical energy reduction factor as a function of surface curvatures under a series of contact angles is computed and shown Figure 1.8. The critical Gibbs free energy decreases with decreasing contact angle (improved wetting). Moreover, a large surface curvature leads to higher reduction of critical energy and enhanced nucleation rate. Against intuition, this implies that larger nanoparticles are more efficient nucleation agents.

The above analysis was first adopted by Shen, Zeng, and Lee (2005) to compare the nucleation efficiency of nanoparticles of different geometry (single-walled carbon nanotubes, carbon nanofibers, and clay nanoparticles) in polystyrene foams. They found that consistent with the theoretical prediction, among the three nanoparticles studied, the single-walled carbon nanotubes exhibited the lowest nucleation efficiency because of the smallest size and curvature. Carbon nanofibers have the highest nucleation efficiency and nanoclays (with flat surface) have an efficiency in between.

Later, Goren et al. (2010) conducted a more systematic study on the size effect using nanoparticles with the same base geometrical shape (spherical). They prepared PMMA silica nanocomposite foams using two nanosilica of different sizes

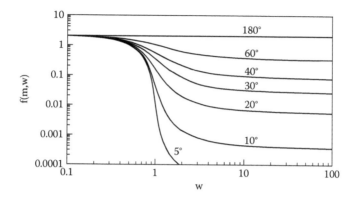

**FIGURE 1.8** Reduction of critical Gibbs free energy for nucleation as functions of surface curvature and contact angle. The numbers atop the curves are the contact angle values used for the construction of the particular curve.

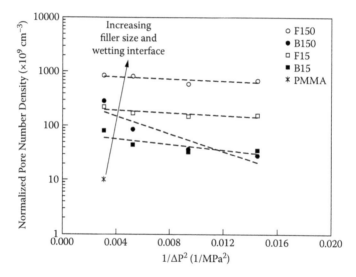

**FIGURE 1.9** Influence of nanoparticle surface chemistry and size on supercritical carbon dioxide processed nanocomposite foam morphology. (Reprinted with permission from Goren K. et al., *Journal of Supercritical Fluids* 2010, 51, 420–427. Copyright 2010, Elsevier.)

(150 nm and 15 nm) and found that reducing the size of nanoscale silica led to decreased nucleation efficiency because of the increased nucleation free energy due to the surface curvature effect. Nevertheless, the reduction in nucleation efficiency was countered by the fact that decreasing the particle size would provide many more nucleation sites, and ultimately higher bubble density. Their study of the particle size on the nucleation efficiency is summarized in Figure 1.9, which also contains a discussion on the surface chemistry effects (Section 1.4.1.3).

For anisotropic nanoparticles, the aspect ratio would play a role in nucleation efficiency and is a direct consequence of the surface curvature effect. In the studies by

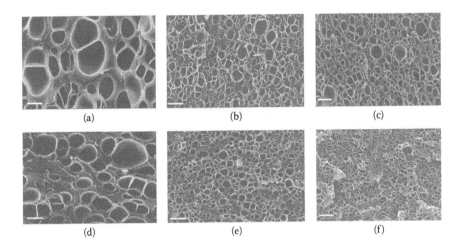

**FIGURE 1.10** SEM micrographs: (a) Neat PMMA. (b) M100 nanocomposites. (c) M20 nanocomposites foamed at 17.9 MPa (top row). (d) Neat PMMA. (e) M100 nanocomposites. (f) M20 nanocomposites foamed at 15.8 MPa (bottom row). Scale bar = 4 μm. (Reprinted with permission from Chen L. et al., *Polymer* 2010, 51, 2368–2375. Copyright 2010, Elsevier.)

Chen, Ozisik, and Schadler (2010) and Chen et al. (2012) multi-walled carbon nanotubes (MWCNTs) with controlled aspect ratio were used to alter the bubble density in PMMA/MWCNTs nanocomposites. It was found that the PMMA nanocomposite foams with shorter MWCNTs (M20) had higher bubble density than those with longer MWCNTs (M100) under the same foaming conditions and CNT concentration, as shown in Figure 1.10. They provided the reasoning based on Fletcher's theory. Both the ends and sidewalls of carbon nanotubes can act as heterogeneous bubble nucleation sites, but the ends (flat surfaces) have larger surface curvatures than the sidewalls (curved surfaces) and therefore are more effective nucleation centers. At the same CNT concentration, short nanotubes would possess more tube ends than long nanotubes and therefore more nucleation sites and higher bubble density.

The effect of particle concentration on the PMMA foam nucleation was investigated (Zeng, 2004). The cell density was found to increase linearly versus clay concentration at low clay concentration, and starts to level off as clay concentration further increases, arguably resulting from the deleterious particle dispersion as the concentration increases.

### 1.4.1.2 Effect of Nanoparticle Dispersion

The effect of nanoparticle dispersion on the foam cell morphology is readily understood by the nucleation theories described above (Equation 1.1). As nanoparticles were better dispersed, higher effective particle concentration ($C_{het}$) would have been achieved at the same nominal concentration. However, dispersion is critically important in obtaining good foam morphology. If the nanoparticle dispersion is not sufficient, nonuniform cellular structure may result (Manninen et al., 2005).

Zeng et al. (2003) prepared PMMA nanoclay nanocomposites with different dispersions: intercalated (PMMA/20A) versus exfoliated (PMMA/MHABS).

The clay dispersion was shown in Figure 1.2. Figure 1.11 shows a comparison between PMMA and PMMA nanocomposite foams. The exfoliated nanocomposites yielded a much higher cell density and smaller cell size than the intercalated nanocomposites.

The PMMA/MHABS nanocomposite foam exhibits dramatically smaller cell size (1.7 μm) and higher cell density ($1.51 \times 10^{11}$ cells/cm$^3$), when compared with PMMA or PMMA/20A (Figure 1.11d). Further investigation revealed that the surface chemistry of the clay nanoparticle played a significantly more important role than the improved nanoparticle dispersion, albeit a substantial increase in the nucleation rate did arise from the latter. This will be discussed in more detail in Section 1.4.1.3.

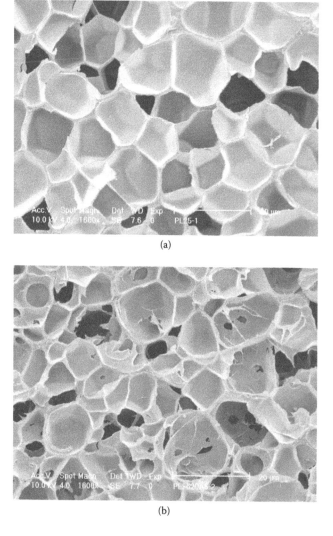

(a)

(b)

**FIGURE 1.11**    SEM micrographs: (a) PMMA. (b) PMMA/5% 20A nanocomposite.

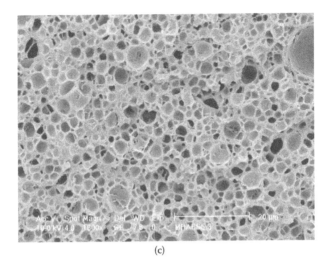

(c)

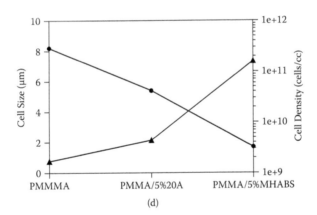

(d)

**FIGURE 1.11 (Continued)** SEM micrographs: (c) PMMA/5% MHABS nanocomposite foams at 120°C and $CO_2$ pressure of 13.8 MPa. (d) Summary of cell size and cell density. (Reprinted with permission from Zeng C. et al., *Advanced Materials* 2003, 15, 1743–1747. Copyright 2003, John Wiley & Sons.)

Yeh et al. (2009) also found that the nature of the dispersion of organoclay plays a vital role in controlling the size of the cell during foaming and in exfoliated nanocomposites; in particular, the individual particles enable a much larger interfacial area between clay particles and the polymer matrix to be used for cell nucleation.

Zeng et al. (2010) studied the effects of carbon nanotube dispersion and PMMA foam cell morphology. They observed that in nanocomposites where CNTs were not well dispersed, the foam exhibited bimodal cell size distribution (Figure 1.12). A smaller amount of big bubbles are distributed in a large number of significantly smaller bubbles. By analysis based on nucleation theory and deliberate experiments, they elucidated that such bimodal cell size distribution resulted from mixed mode

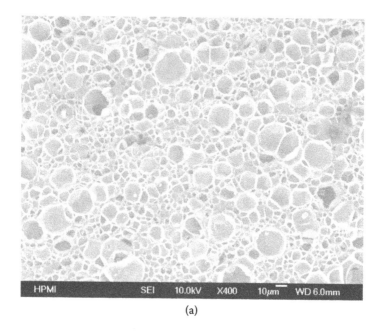

(a)

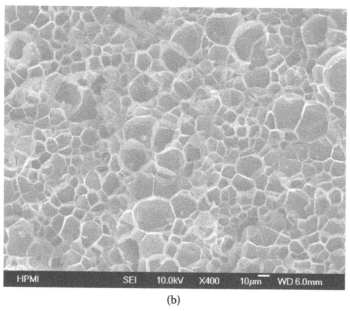

(b)

**FIGURE 1.12** Bimodal cell size distribution observed in PMMA-1wt% CNT nanocomposite foams prepared at 13.8 MPa and different temperatures: (a) 100°C; (b) 120°C.

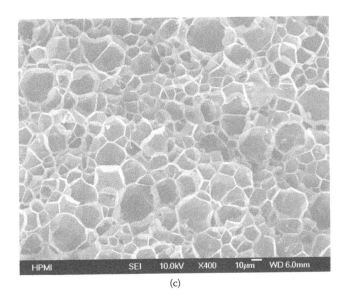

(c)

**FIGURE 1.12 (*Continued*)**    Bimodal cell size distribution observed in PMMA-1wt% CNT nanocomposite foams prepared at 13.8 MPa and different temperatures: (c) 140°C. Scale bars: 10 μm in all micrographs. (Reprinted with permission from Zeng C. et al., *Polymer* 2010, 51, 655–664. Copyright 2010, Elsevier.)

nucleation, where heterogeneous nucleation prevailed in the CNT's rich region and homogeneous nucleation readily proceeded in the polymer rich region. The former led to the large amount of smaller bubbles while the latter resulted in the large bubbles. Furthermore, the degree of nanoparticle dispersion necessary to achieve uniform cell size distribution is closely associated with the foaming regime and overall nucleation density. The higher the potential nucleation rate, the more critical the uniformity of nanoparticle dispersion is. This is discussed in detail in Chen et al. (2011).

### 1.4.1.3    Effect of Surface Chemistry of Nanoparticles
The effects of the surface chemistry of the nanoparticles are twofold. First, the surface functionalization typically led to improved polymer–nanoparticle interaction and improved nanoparticle dispersion. This has been discussed extensively in previous sections. Second, the change of the surface chemistry would profoundly change the nucleation process by affecting the critical nucleation of free energy. By introducing surface moieties that affine to the foaming agent, for example, carbon dioxide, to reduce the contact angle; or chemicals that have intrinsically low surface tension, the critical nucleation free energy would be reduced. As the free energy affects the nucleation rate in an exponential manner (Equation 1.1), this would result in a significantly enhanced nucleation rate.

This concept was first demonstrated by Zeng et al. (2003) and Zeng (2004) in their study of PMMA and polystyrene (PS) clay nanocomposite foams. When a PMMA molecule was polymerized from the nanoclay surface, it not only led to the exfoliation of the nanoclay but also anchored the PMMA molecules on the clay surface. The favorable

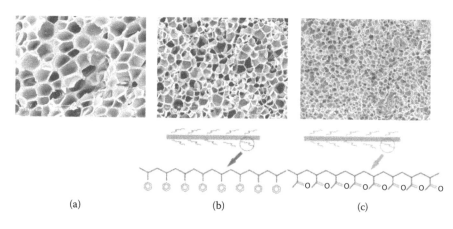

(a)                                         (b)                                         (c)

**FIGURE 1.13**   An example of tuning the nanoclay surface chemistry to control the cell morphology in PS foams. (a) Pure PS foam as a reference; clays were exfoliated in both (b) (with 5%) and (c) with (2%) but with different surface chemistry. The surface was covered with PS in (b) and PMMA in (c). Due to the reduction of free nucleation energy and higher nucleation rate, the foam in (c) has a significantly higher cell density than (b) despite the lower clay concentration. (Reprinted with permission from Zeng C. et al., *Advanced Materials* 2003, 15, 1743–1747. Copyright 2003, John Wiley & Sons; Reprinted with permission from Lee L. J. et al., *Composites Science and Technology* 2005, 65, 2344–2363. Copyright 2005, Elsevier.)

interaction of the carbonyl group (on PMMA) and carbon dioxide (Kazarian et al., 1996; Kazarian, 2002) resulted in a reduction of the contact angle and significant increases in the cell nucleation rate. With 5% nanoclays (MHABS) about two order of magnitude increases in cell density were achieved in the PMMA nanocomposite foam. By comparison, when PS molecules are anchored on the nanoclay surface, while the same exfoliation was achieved, only a five times increase of cell density (in PS) was observed. When PMMA-grafted nanoclay was exfoliated in PS, a much higher cell density and smaller cell size was achieved at a lower clay concentration (2 wt%), further validating this concept. The cell morphology and nanoparticle surface difference is shown in Figure 1.13.

Later, Goren et al. (2010) conducted a more comprehensive investigation of the effects of nanoparticle surface chemistry on nucleation during silica/PMMA nanocomposites using a series of foaming pressures and the results are summarized in Figure 1.9. Reduction of surface tension via fluorination (resulting from the strong interaction affinity between $CO_2$ and fluorinated compounds) (Folk, DeSimone, and Samulski, 2001; Sarbu, Styranec, and Beckman, 2000) of the silica nanoparticle surface led to decreased cell size without changing the degree of silica aggregation and overall foam density. Note this figure also includes the studies on the particle size effect described earlier.

The effects of CNT's surface chemistry on the bubble density were also studied (Chen et al., 2012). In their study, the oxidized CNTs (M20 and M100) were functionalized by grafting with glycidyl phenyl ether (GPE) (Figure 1.14a). The GPE functionalized CNTs (*P*20 and *P*100) would have the same aspect ratio as the parent oxidized CNTs. They found that at the same nanotube concentration, the nanocomposite foams with GPE grafted CNTs had a bubble density several times higher than that of nanocomposite foams with oxidized CNTs under a series of foaming

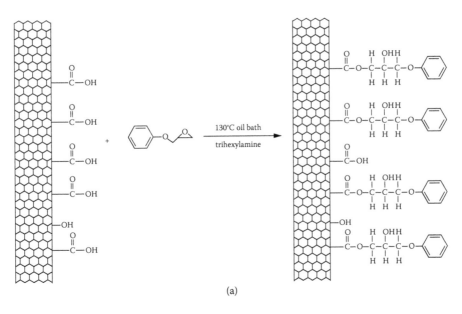

(a)

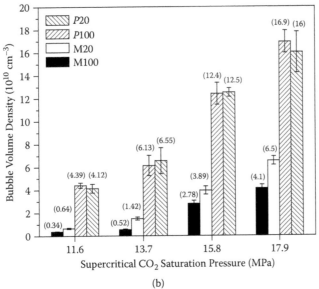

(b)

**FIGURE 1.14** Effects of MWCNT surface functionalization on the bubble density of PMMA CNT nanocomposite foams. (a) Surface functionalization scheme, in which oxidized CNTs (M20 and M100) were surface grafted with glycidyl phenyl ether (P20 and P100). 20, 100 in the notation refer to the CNT aspect ratio. (b) A comparison of the bubble nucleation density of M100, P100, M20, and P20 nanocomposites foamed, saturated at various pressures, and foamed at 65°C for 5 minutes. Error bars represent the standard deviations of bubble nucleation density data of four samples under each condition. (Reprinted with permission from Chen L. et al., *Composites Science and Technology* 2012, 72, 190–196. Copyright 2012, Elsevier.)

conditions (Figure 1.14b). They further asserted that after GPE function, the surface chemistry became the dominant factor that governed the bubble density, while the effect of the nanotube aspect was diminished.

### 1.4.1.4   Effect of the Nanoparticle on Matrix Rigidity

It is well known that well dispersed nanoparticles often lead to significant increases in the viscosity, melt strength, and rigidity of the polymer matrix. These would have profound influences on bubble expansion and foam morphology. While the enhanced melt strength has been observed in improving foaming of low melt strength polymer such as polypropylene (PP) (Okamoto et al., 2001), excessive matrix rigidity enabled by a high concentration of well-dispersed nanoparticles and exacerbated by low foaming temperature would lead to complete inhibition of foaming. This has been observed in the retrograde foaming of PMMA-CNT nanocomposite foams and has been discussed in Section 1.3.3.

### 1.4.2   Properties

In ever increasing studies, nanocomposite foams show potential improvement in a wide range of properties, for example, mechanical, electrical, and thermal properties, when compared with virgin nanocomposites or conventional foams.

Zeng et al. (2013) studied the tensile properties of PMMA/MWCNT nanocomposite foams, which were determined by the convoluted effects of CNT dispersion, polymer–CNT interaction, and foam structure differences. Whereas foams containing poorly dispersed CNTs with weak polymer–CNT interaction showed reduction in both tensile strength and modulus, simultaneous improvement in tensile strength, modulus, and elongation at break were observed in nanocomposite foams in which CNTs were well dispersed and had good affinity to the polymer matrix. Nanocomposite foam with concurrent increases in tensile strength (~40%), tensile modulus (~60%), and strain at break (~70%) was successfully prepared with the use of 0.5% functionalized CNTs that were well dispersed. The foam showed a ductile failure under tension that involved extensive pore deformation and collapsing, and formation and coalescence of microvoids that were largely responsible for the significantly improved tensile toughness (Figure 1.15). It shall be noted that while promising, the exact physical origins of these multiple energy dissipation mechanisms (Sun et al., 2002) are still not well understood and warrant further investigation in the future.

Kynard (2011) investigated the energy dissipation capabilities of the PMMA nanocomposite foams by examining the compressive toughness. The toughness was obtained by integrating the compression stress–strain curve and used as an indication for energy absorbing capabilities. The 0.5% CNT nanocomposite foam showed improvement in energy absorption while the capabilities decreased when CNT concentration increased.

Chen, Schadler, and Ozisik (2011) investigated the compressive properties of PMMA/MWCNT nanocomposite foams. As shown in Figure 1.16, nanocomposite foams have greater modulus and collapse strength than the neat PMMA foam across the foam density range studied, and the effects were more prominent for nanotubes with higher aspect ratios. The addition of only 1% of MWCNTs (F-C100 with an aspect ratio 100) led to 82% increase in the Young's modulus and 104% increase in the collapse

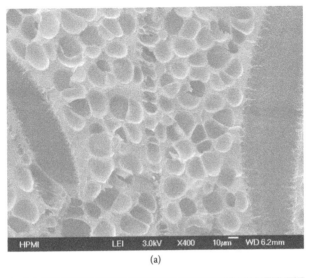

(a)

(b)

**FIGURE 1.15** Failure surfaces of PMMA (a) and (b) PMMA-0.5% CNT nanocomposite foams. The PMMA foam showed brittle failure while the nanocomposite foam exhibited ductile failure with extensive deformation contributing to the improvement in tensile toughness. (Reprinted with permission from Zeng C. et al., *Composites Science and Technology* 2013, 82, 29–37. Copyright 2013, Elsevier.)

strength (at a relative density of 0.5). Both modulus and collapse strength decreased with decreasing relative density. The authors argued that addition of CNTs influence the foam properties by (i) improving the compressive properties of the matrix and (ii) reducing the bubble size of the nanocomposite foams. A model taking into account the bubble size effect was derived for predicting the compressive properties of the foams. Yeh et al. (2011) also investigated the compressive properties of PMMA-CNT

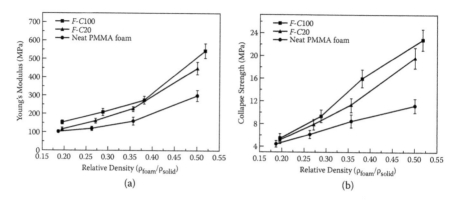

**FIGURE 1.16** Plots showing the change of (a) Young's modulus and (b) collapse strength of nanocomposite foams and neat PMMA foams with relative density. The error bars represent the standard deviations of Young's modulus/collapse strength of five specimens for each sample. (Reprinted with permission from Chen L., Schadler L. S., and Ozisik R. *Polymer* 2011, 52, 2899–909. Copyright 2011, Elsevier.)

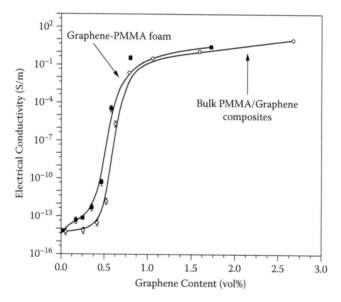

**FIGURE 1.17** Plots of electrical conductivity versus graphene content for PMMA/graphene bulk nanocomposites and microcellular foams. (Reprinted with permission from Zhang H. et al., *ACS Applied Materials & Interfaces* 2011, 3, 918–924. Copyright 2011, American Chemical Society.)

nanocomposite foams and observed 160% increase in compressive modulus with the addition of 0.3 wt% carboxyl-multi-walled carbon nanotubes (c-MWNTs).

Properties other than mechanical properties were also studied. Zhang et al. (2011) investigated the effect of graphene content on the electrical conductivity of PMMA/graphene nanocomposite foams (Figure 1.17) and found that the percolation

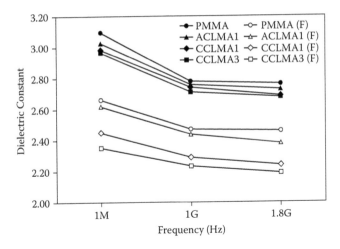

**FIGURE 1.18** Dielectric constant of PMMA clay nanocomposite and nanocomposite foams with two kinds of nanoclays (CCLMA and ACLMA) at various frequencies under room temperature. Number in legend is the clay concentration; (F) indicates foam. (Reprinted with permission from Yeh J. et al., *Materials Chemistry and Physics* 2009, 115, 744–750. Copyright 2009, Elsevier.)

threshold of the foams (with an expansion ratio of ca. 2) shifted to lower graphene content compared with that of the bulk nanocomposites.

Yeh et al. (2009) investigated the effect of nanoclay on the dielectric and thermal transport properties of PMMA nanocomposite foams. As shown in Figure 1.18, the nanocomposite foams showed lower dielectric constants than the neat PMMA foam. And the effect is more prominent when the clay nanoparticles were better dispersed (CCLMA clay) and when the clay concentration was increased. The effect on thermal conductivity (Figure 1.19) was slightly more complicated. While the nanocomposite foams with better dispersion, that is, CCLMA nanocomposites with an exfoliated–intercalated mixed morphology, showed a decrease in thermal conductivity, the thermal conductivity of the intercalated ACLMA nanocomposite foam was higher than that of neat PMMA foam. They have also prepared PMMA MWCNT nanocomposite foams and measured their insulation property. Interestingly, they noticed a decrease in both dielectric constant (22.6%) and thermal conductivity (19.7%) in the nanocomposite foams with 0.3 wt% carboxyl-multi-walled carbon nanotubes (c-MWNTs).

## 1.5 APPLICATIONS

As discussed earlier, one of the driving forces in nanocomposite foam research is to leverage the reinforcement efficiency of the various nanoparticles, coupled with the enabled morphological control, to produce materials that possess higher mechanical properties. These have been discussed in the previous sections. In the following sections, we will discuss *unconventional* applications of the PMMA nanocomposite foams enabled by the unique properties of the nanoparticles.

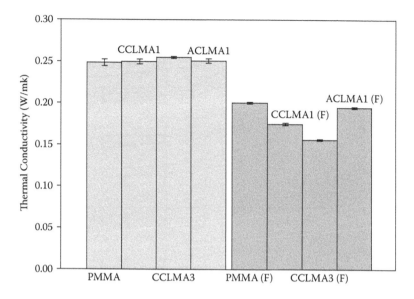

**FIGURE 1.19** Thermal conductivity of PMMA clay nanocomposite and nanocomposite foams with two kinds of nanoclays (CCLMA and ACLMA) measured at room temperature. Number in legend is the clay concentration; (F) indicates foam. (Reprinted with permission from Yeh J. et al., *Materials Chemistry and Physics* 2009, 115, 744–750. Copyright 2009, Elsevier.)

### 1.5.1 Electromagnetic Interference Shielding

Electromagnetic interference (EMI) is problematic because it disturbs the normal function of electronics and may cause irradiative damage to the human body (Li et al., 2008). Conventional EMI shielding materials are typically metals and their composites. Because of the high conductivity and high dielectric constant, they have a high shielding effectiveness. But they are heavy and have poor corrosion resistance (Yang et al., 2005).

Nanocomposite foams with high-conductivity nanoparticles such as carbon nanofibers, carbon nanotubes, or graphene have a potential for effective EMI shielding materials with light weight and significantly improved corrosion resistance and environmental durability. The EMI shielding efficiency depends on the electrical conductivity and bubble density and interconnectivity, which, in turn, depends on the properties of the nanoparticles, dispersion, and foaming conditions. In addition, the main EMI shielding mechanism of the solid materials is reflection and is not always desired because of the potential damage from internal reflection. The nanocomposite foams, on the other hand, utilize both reflection and absorption (Thomassin et al., 2008). In the latter, multiple reflections of the electromagnetic wave occur within the cavities of the foams resulting in improved efficiency and better protection.

PMMA/graphene foams were prepared and the EMI shielding property was investigated (Zhang et al., 2011). It was found that with a low graphene loading of 1.8 vol%, the nanocomposite foam exhibited both high conductivity of 3.11 $Sm^{-1}$ and good EMI shielding efficiency of 13–19 dB at frequencies from 8 to 12 GHz, which is close to the target value of EMI shielding efficiency required for practical applications (ca. 20 dB). The EMI shielding efficiency is mainly attributed to absorption rather than reflection in the investigated frequency range. In the PMMA/ graphene foam, incident microwaves entering the microcellular foam were reflected and scattered many times between cell–matrix interfaces and the graphene sheets and had difficulty escaping from the material until they were absorbed (Shen et al., 2012). Their results are summarized in Figure 1.20 (Zhang et al., 2011).

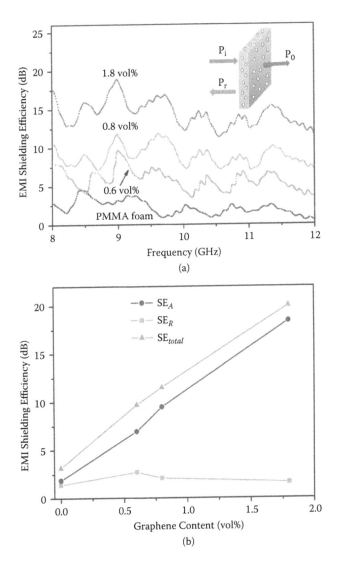

**FIGURE 1.20** PMMA graphene nanocomposite foams for EMI shielding; (a) EMI shielding efficiency of graphene–PMMA nanocomposite microcellular foams with different contents of graphene sheets. (b) The comparison of $SE_{total}$, microwave absorption ($SE_A$), and microwave reflection ($SE_R$) at 9 GHz. (Reprinted with permission from Zhang H. et al., *ACS Applied Materials & Interfaces* 2011, 3, 918–924. Copyright 2011, American Chemical Society.)

## 1.5.2  Tissue Engineering Applications

Another *unconventional* application of PMMA nanocomposite foams is in the area of tissue engineering, which focuses on developing artificial tissues and/or organs by unifying biology, engineering, and materials (Lanza, Langer, and Vacanti, 2000). A scaffold is a three-dimensional porous construct used as a support structure allowing biological cells to adhere, proliferate, and differentiate to form a healthy tissue (Liebschner and Wettergreen, 2003). The scaffold plays a critical role in that it affects many aspects of the tissue growth, for example, cell seeding, cell migration, matrix deposition, vascularization, and mass transport of nutrients to and from the cells (Yang et al., 2001). Scaffolds need to be biocompatible, radiolucent, easily formable, nonallergic, noncarcinogenic, and mechanically stable having a hierarchical macroporous framework (Hui, Leung, and Sher, 1996).

PMMA is a biocompatible and FDA-approved orthopedic material and has been extensively used in bone cements to fix load-bearing prosthetic components in total joint replacement surgery (Ohgaki and Yamashita, 2003). It has been reported that the integration of hydroxyapatite (HA) nanoparticles with PMMA can provide bioactivity coupled with mechanical stability (McManus et al., 2005). More recently, Sinha, Guha, and Sinha (2010) developed a method to prepare the PMMA/HA nanocomposite foam with a honeycomb structure. The morphology is shown in Figure 1.21. One mm-sized macropores were uniformly distributed and symmetrically surrounded by 150–200 μm-sized pores (Figure 1.21a), and higher magnification view confirmed the interconnectivity of the macropores (Figure 1.21b) and presence of 2–30 μm-sized micropores on the struts of the framework (Figure 1.21c). Though still at a very early stage, the initial results do suggest that such scaffolds promote mesenchymal stem cell adhesion (Figure 1.22).

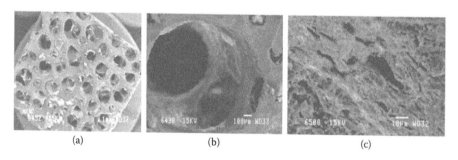

(a)                              (b)                              (c)

**FIGURE 1.21**   (a) SEM micrograph of HA-PMMA composite. (b) Magnified SEM micrograph manifesting interconnectivity of the macropores in HA-PMMA composite. (c) Magnified SEM images confirming microporous structure of the strut. (Reprinted with permission from Sinha S. et al., *Materials Science & Engineering, C: Materials for Biological Applications* 2010, 30, 6, 873–877. Copyright 2010, Elsevier.)

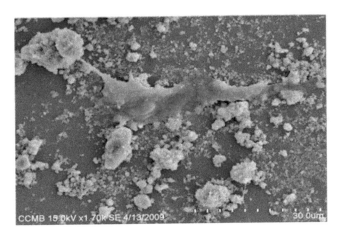

**FIGURE 1.22**    Adhesion of mesenchymal stem cells on the surface of HA-PMMA composite. (Reprinted with permission from Sinha S. et al., *Materials Science & Engineering, C: Materials for Biological Applications* 2010, 30, 6, 873–877. Copyright 2010, Elsevier.)

## 1.6    CONCLUSIONS AND OUTLOOK

PMMA foam is a versatile polymer foam that possesses many desirable properties and can be used in many applications. Along with the numerous choices of nanoparticles, a myriad of PMMA nanocomposite foams can be produced to suit a particular application. Nevertheless, similar to other polymer nanocomposite foams, tremendous challenges exist before the potential can be unleashed. These include the understanding of the structure properties of the nanocomposite foams, and the convoluted role of the nanoparticle surface chemistry, dispersion, polymer–nanoparticle interaction, and foaming processing conditions that govern the cellular structure formation.

## REFERENCES

Abraham F. F., *Advances in Theoretical Chemistry, Suppl. 1: Homogeneous Nucleation Theory. The Pretransition Theory of Vapor Condensation*, 1974, Academic Press: New York.

Ashby M. F., Ferreira P. J., and Schodek D. L., *Nanomaterials, Nanotechnologies and Design: An Introduction for Engineers and Architects*, 2009, Oxford: Elsevier, p.177.

Bauhofer W. and Kovacs J. Z., A review and analysis of electrical percolation in carbon nanotube polymer composites. *Compos Sci Technol* 2009, 69, 1486–1498.

Blumstein A., Polymerization of adsorbed monolayers. I. Preparation of clay-polymer complex. *Journal of Polymer Science Part A: General Papers* 1965, 3, 2653–2664.

Burda C., Chen X., Narayanan R., and Ei-Sayed M. A., Chemistry and properties of nanocrystals of different shapes. *Chemistry Review* 2005, 105, 1025–1102.

Chatterjee A., Properties improvement of PMMA using nano TiO2, *Journal of Applied Polymer Science* 2010, 118, 2890–2897.

Chen L., Goren B. K., Ozisik R., and Schadler L. S., Controlling bubble density in MWNT/polymer nanocomposite foams by MWNT surface modification. *Composites Science and Technology* 2012, 72, 190–196.

Chen Z., Kynard K., Zeng C., Zhang, C., and Wang B., Foaming of polymer carbon nanotube nanocomposite from the retrograde phase. *SPE ANTEC* 2011, 69, 2678–2682.

Chen L., Ozisik R., and Schadler L. S., The influence of carbon nanotube aspect ratio on the foam morphology of MWNT/PMMA nanocomposite foams. *Polymer* 2010, 51, 2368–2375.

Chen L., Schadler L. S., and Ozisik R. An experimental and theoretical investigation of the compressive properties of multi-walled carbon nanotube/poly(methyl methacrylate) nanocomposite foams. *Polymer* 2011, 52, 2899–2909.

Chen M., Wu L., Zhou S., and You B., Synthesis of raspberry-like PMMA/SiO$_2$ nanocomposite particles via a surfactant-free method. *Macromolecules* 2004, 37, 9613–9619.

Condo P. D., Paul D. R., and Johnston, K. P., Glass transition of polymers with compressed fluids diluents: Type II and III behavior. *Macromolecules* 1994, 27, 365–371.

Cooper A. I., Polymer synthesis and processing using supercritical carbon dioxide. *Journal of Materials Chemistry* 2000, 10, 207–234.

Darder M., Aranda P., Ferrer M. L., Gutiérrez M. C., del Monte F., and Ruiz-Hitzky E., Progress in bionanocomposite and bioinspired foams. *Advanced Materials* 2011, 23, 5262–5267.

Du F., Fischer J. E., and Winey K. I. Coagulation method for preparing single-walled carbon nanotube/poly(methyl methacrylate) composites and their modulus, electrical conductivity, and thermal stability. *Journal of Polymer Science Part B: Polymer Physics* 2003, 41(24), 3333–3338.

Fletcher N. H., Size effect in heterogeneous nucleation. *Journal of Chemical Physics* 1958, 28, 572–576.

Folk S. L., DeSimone J. M., and Samulski E. T., Cationic poly(dimethylsiloxane) surfactants: Synthesis, characterization, and aggregation behavior in dense carbon dioxide, fluorinated, and silicon-containing solvents. *Polymer* 2001, 42, 231–232.

Fu J. and Naguib H. E., Effect of nanoclay on the mechanical properties of PMMA/clay nanocomposite foams. *Journal of Cellular Plastics* 2006, 42, 325–342.

Goren K., Chen L., Schadler L. S., and Ozisik R., Influence of nanoparticle surface chemistry and size on supercritical carbon dioxide processed nanocomposite foam morphology. *Journal of Supercritical Fluids* 2010, 51, 420–427.

Gorga R. E. and Cohen R. E., Toughness enhancements in poly(methyl methacrylate) by addition of oriented multiwall carbon nanotubes. *Journal of Polymer Science Part B: PolymerPhysics* 2004, 42, 2690–2702.

Han X., Zeng C., Lee L. J., Koelling K. W., and Tomasko D. L., Extursion of polystyrene nanocomposite foams with supercritical CO$_2$. *Polymer Engineering and Science* 2003, 43, 1261–1275.

Handa Y. P. and Zhang Z., A new technique for measuring retrograde vitrification in polymergas systems and for making ultramicrocellular foams from the retrograde phase. *Journal of Polymer Science: Part B: Polymer Physics* 2000, 38, 716–725.

Handa Y. P., Zhang, Z., and Wong, B., Solubility, diffusivity and retrograde vitrification in PMMA-CO$_2$, and development of sub-microcellular structures. *Cellular Polymers* 2001, 20, 1–16.

Huang X. and Brittain W. J., Synthesis and characterization of PMMA nanocomposites by suspension and emulsion polymerization. *Macromolecules* 2001, 34, 3225–3260.

Hui P. W., Leung P. C., and Sher A., Fluid conductance of cancellous bone graft as a predictor for graft-host interface healing. *Journal of Biomechanics* 1996, 29, 123–132.

Ibeh C. C. and Bubacz M., Current trends in nanocomposite foams. *Journal of Cellular Plastics* 2008, 44, 493–515.

Jo C., Fu J., and Naguib H. E., Constitute modeling for intercalated PMMA/clay nano composite foams. *Polymer Engineering and Science* 2006, 46, 1787–1796.

Johnston K. P. and Shah P. S., Materials science—Making nanoscale materials with supercritical fluids. *Science* 2004, 303, 482–483.

Kashiwagi T., Fagan J., Leigh S. D., Obrzut J., Du F., Lin-Gibson S., Mu M., Winey K. I., and Haggenmueller R., Relationship between dispersion metric and properties of PMMA/SWNT nanocomposites. *Polymer* 2007, 48, 16, 4855–4866.

Kazarian, S. G., Polymers and supercritical fluids: Opportunity for vibration spectroscopy. *Macromolecular Symposia* 2002, 184, 215–228.

Kazarian, S. G., Vincent, M. F., Bright, F. V., Liotta, C. L., and Eckert, C. A., Specific intermolecular interaction of carbon dioxide with polymers. *Journal of the American Chemical Society* 1996, 118, 1729–1736.

Kumar S., Jog J. P., and Natarajan U., Preparation and characterization of poly (methylmethacrylate)-clay nanocomposites via melt intercalation: The effect of organoclay on the structure and thermal properties. *Journal of Applied Polymer Science* 2003, 89, 1186–1194.

Kynard K., PMMA carbon nanotube nanocomposite foams for energy dissipation, M.S. Thesis, Florida State University, 2011.

Laaksonen A., Talanquer V., and Oxtoby D. W., Nucleation: Measurements, theory, and atmospheric applications. *Annual Review of Physical Chemistry* 1995, 46, 489–524.

Lanza, R., Langer R., and Vacanti, J. P., *Principles of Tissue Engineering*, 2000, Academic Press, New York.

Lee D. C. and Jang L. W., Preparation and characterization of PMMA-clay hybrid by emulsion polymerization. *Journal of Applied Polymer Science* 1996, 61, 1117–1122.

Lee L. J., Zeng C. C., Cao X., Han X. M., Shen J., and Xu G. J., Polymer nanocomposite foams. *Composites Science and Technology* 2005, 65, 2344–2363.

Li Y., Chen C., Zhong S., Ni Y., and Huang J., Electrical conductivity and electromagnetic interference shielding characteristics of multiwalled carbon nanotube filled polyacrylate composite films. *Applied Surface Science* 2008, 254, 5766–5771.

Liebschner M. A. K. and Wettergreen M. A., Optimization of bone scaffolding for load bearing applications. In Ashammakhi N. and Ferretti P. (eds.), *Topics in Tissue Engineering*, 2003, 1.

Luo J. Inorganic-organic nanocomposites formed using porous ceramic particles, Ph.D. diss., The Ohio State University, 1998.

Manninen A. R., Naguib H. E., Nawby A. V., Liao X., and Day M., The effect of clay content on PMMA-clay nanocomposites. *Cellular Polymers* 2005, 24, 49–70.

Martini-Vvedensky, J. E., Waldman, F. A. et al., The production and analysis of microcellular thermoplastic foams. *SPE ANTEC* 1982, 28, 674–676.

McManus A. J., Doemus R. H., Siegel R. W., and Bizios R., Evaluation of cytocompatibility and bending modulus of nanoceramic/polymer composites. *Journal of Biomedical Materials Research Part A* 2005, 72, 98–106.

Moniruzzaman M. and Winey K. I., Polymer nanocomposites containing carbon nanotubes. *Macromolecules* 2006, 39, 5194–5205.

Ohgaki M. and Yamashita K. J., Preparation of polymethylmethacrylate-reinforced functionally graded hydroxyapatite composites. *Journal of the American Ceramic Society* 2003, 86, 1440–1442.

Okamoto M., Nam P. H., Maiti P., Kotaka T., Nakayama T., Takada M., Ohshima M., Usuki A., Hasegawa N., and Okamoto H., Biaxial flow-induced alignment of silicate layers in polypropylene/clay nanocomposite foam. *Nano Letters* 2001, 1, 503–505.

Ramanathan T., Stankovich S., Dikin D. A., Liu H., Shen H., Nguyen S. T., and Brinson L. C., Graphitic nanoparticles in PMMA nanocomposites: An investigation of particle size and dispersion and their influence on nanocomposite properties. *Journal of Polymer Science: Part B: Polymer Physics* 2007, 45, 2097–2112.

Sarbu T., Styranec T., and Beckman E. J., Non-fluorous polymers with very high solubility in supercritical $CO_2$ down to low pressures. *Nature* 2000, 405, 165–168.

Shen J., Zeng C., and Lee L. J., Synthesis of polystyrene-carbon nanofibers nanocomposite foams. *Polymer* 2005, 46, 5218–5224.

Shen B., Zhai W., Lu, D., Zheng W., and Yan Q., Fabrication of microcellular polymer/ graphene nanocomposite foams. *Polymer International* 2012, 61, 1693–1702.

Sinha S., Guha A., and Sinha A., Macroporous hybrid frameworks for bone graft substitute. *Materials Science & Engineering, C: Materials for Biological Applications* 2010, 30, 6, 873–877.

Siripurapu S., DeSimone J. M., Khan S. A., and Spontak R. J., Controlled foaming of polymer films through restricted surface diffusion and the addition of nanosilica particles or $CO_2$-philic surfactants. *Macromolecules* 2005, 38, 2271–2280.

Suhr J. and Koratkar N. A., Energy dissipation in carbon nanotube composites: A review. *Journal of Materials Science* 2008, 43, 4370–4382.

Sun L., Gibson R. F., Gordaninejad F., and Suhr J., Energy absorption capability of nanocomposites: A review. *Composites Science and Technology* 2009, 69, 2392–2409.

Sun H., Sur G. S., and Mark J. E., Microcellular foams from polyethersulfone and polyphenylsulfone-preparation and mechanical properties. *Europe Polymer Journal* 2002, 38, 2373–2381.

Thomassin J., Pagnoulle C., Bednarz L., Huynen I., Jerome R., and Detrembleur C., Foams of polycaprolactone/MWNT nanocomposites for efficient EMI reduction. *Journal of Material Chemistry* 2008, 18, 792–796.

Thomassin J., Vuluga D., Alexandre M., Jerome C., Molenberg I., Huynen I., and Detrembleur C., A convenient rout for the dispersion of carbon nanotubes in polymers: Application to the preparation of electromagnetic interference (EMI) absorbers. *Polymer* 2012, 53, 169–174.

Wang Y. and Guo J., Melt compounding of PMMA/clay nanocomposites with styrene-maleic anhydride copolymers: Effect of copolymer type on thermal, mechanical and dielectric properties. *Polymer Composited* 2010, 31, 596–603.

Wang, H., Zeng, C., Elkovitch, M., Lee, L. J., and Koelling, K. W., Processing and properties of polymeric nano-composites. *Polymer Engineering and Science* 2001, 41, 2036–2046.

Wang D., Zhu J., Yao Q., and Wilkie C. A., A comparision of various methods for the preparation of polystyrene and poly(methyl methacrylate) clay nanocomposites. *Chemistry of Materials* 2002, 14, 3837–3843.

Yang Y. L., Gupta M. C., Dudley K. L., and Lawrence R. W., A comparative study of EMI shielding properties of carbon nanofiber and multi-walled carbon nanotube filled polymer composites. *Journal of Nanoscience Nanotechnol* 2005, 5, 927–931.

Yang S., Leong K. F., Du Z., and Chua C. K., The design of scaffolds for use in tissue engineering. Part I. Traditional factors. *Tissue Engineering* 2001, 7, 679–689.

Yang F. and Nelson G. L., PMMA/silica nanocomposite studies: Synthesis and properties. *Journal of Applied Polymer Science* 2004, 91, 3844–3850.

Yeh J., Chang K., Peng C., Lai M., Hung C., Hsu S., Hwang S., Lin H., and Lin H., Effect of dispersion capability of organoclay on cellular structure and physical properties of PMMA/clay nanocomposite foams. *Materials Chemistry and Physics* 2009, 115, 744–750.

Yeh J., Chang K., Peng C., Lai M., Hwang S., Lin H., and Liou S., Enhancement in insulation and mechanical properties of PMMA nanocomposite foams infused with multi-walled carbon nanotubes. *Journal of Nanoscience and Nanotechnology* 2011, 11, 6757–6764.

Zeng C., Polymer nanocomposites: Synthesis, structure and processing, Ph.D. diss., The Ohio State University, 2004.

Zeng C., Han X., Lee L. J., Koelling K. W., and Tomasko D. L., Polymer–clay nanocomposite foams prepared using carbon dioxide. *Advanced Materials* 2003, 15, 1743–1747.

Zeng C., Hossieny N., Zhang C., and Wang B., Synthesis and processing of PMMA carbon nanotube nanocomposite foams. *Polymer* 2010, 51, 655–664.

Zeng C., Hossieny N., Zhang C., Wang B., and Walsh S., Morphology and tensile properties of PMMA carbon nanotube nanocomposite foams. *Composites Science and Technology* 2013, 82, 29–37.

Zeng C. and Lee L. J., Poly(methyl methacrylate) and polystyrene/clay nanocomposites prepared by *in-situ* polymerization. *Macromolecules* 2001, 34, 4098–4103.

Zhang H., Yan Q., Zheng W., He Z., and Yu Z., Tough graphene-polymer microcellular foams for electromagnetic interference shielding. *ACS Applied Materials & Interfaces* 2011, 3, 918–924.

# 2 Nanotoughening and Microtoughening of Polymer Syntactic Foams

*Xiao Hu, Ming Liu, Erwin M. Wouterson, and Liying Zhang*

## CONTENTS

## SYNOPSIS

The strategies of toughening polymer foams, particularly rigid polymer syntactic foams, are discussed. The discussion begins with a general introduction of toughness and the toughening mechanism in brittle polymer systems before focusing on two of the very successful and effective toughening strategies of thermoset/hollow sphere syntactic foams using microfibers and nanoparticles. Finally, the possible mechanisms for the substantial toughening effect using even a small amount of micro- or nanoparticles in the syntactic foams are proposed and analyzed. It is believed that such an in-depth understanding should enable the formulation of suitable toughening methods for different polymer foam systems.

## 2.1  INTRODUCTION

Most of the polymeric materials, especially thermosets used for structural and composite fabrications, are of high cross-linking density in order to achieve high modulus and stiffness; however, these properties are accompanied by brittleness and low fracture toughness. The fracture toughness refers to the ability of a crack-containing material to resist fracture, and is one of the most important material properties for application design considerations. The introduction of a second phase of well-dispersed particles to initiate localized energy-absorbing mechanisms in the fracture zone, hence suppressing crack propagation, was identified as an effective approach. This strategy was first used on thermoplastics and was found to achieve excellent improvements in impact and fracture resistance. The same strategy was later employed for the toughening of rigid polymer foams to expand the potential applications of these brittle materials to various engineering fields.

The fracture toughness of materials can be quantified by the critical stress intensity factor, $K_{Ic}$, and the critical strain energy release rate, $G_{Ic}$. $K_{Ic}$ means that the stress near the crack tip reaches the point at which the crack will start to grow. $G_{Ic}$ is the amount of the energy required to create the new surface during crack propagation (Rutz and Berg, 2010). The most widely used fracture toughness test configurations are the single-edge notch bend (SENB or three-point bend), as shown in Figure 2.1, and the compact tension (CT) specimens (Wouterson et al., 2005, 2007a,b). The method for determining $K_{Ic}$ and $G_{Ic}$ depends on the test specimen geometry.

For a single-edge notched bend (SENB) specimen (Wouterson et al., 2005, 2007a,b):

$$K_{IC} = Y\frac{3PS\sqrt{a}}{2BW^2} \tag{2.1}$$

**FIGURE 2.1**  Typical test setup of a three-point bending test of SENB specimen.

$$Y = 1.93 - 3.07\left(\frac{a}{w}\right) + 14.53\left(\frac{a}{w}\right)^2 - 25.11\left(\frac{a}{w}\right)^3 + 25.80\left(\frac{a}{w}\right)^4 \quad (2.2)$$

$$G_{IC} = \frac{K_{Ic}^2}{E_t}(1 - \upsilon^2) \quad (2.3)$$

For a compact tension (CT) specimen (Kinloch et al., 1983; Kinloch and Young, 1983),

$$K_{IC} = \frac{PY}{B\sqrt{W}} \quad (2.4)$$

$$Y = 29.6\left(\frac{a}{w}\right)^{\frac{1}{2}} - 185.5\left(\frac{a}{w}\right)^{\frac{3}{2}} + 655.7\left(\frac{a}{w}\right)^{\frac{5}{2}} - 1017\left(\frac{a}{w}\right)^{\frac{7}{2}} + 638.9\left(\frac{a}{w}\right)^{\frac{9}{2}} \quad (2.5)$$

Where for both cases, $Y$ is the geometry correction factor, $P$ is the peak load at the onset of crack growth in a linear elastic fracture, $B$ is the specimen thickness, $W$ is the width of the specimen, $S$ is the support span, $a$ is the crack length, and E is the Young's modulus. It is noted that for SENB samples, the notch length should be between 0.5 and 0.6 times the specimen width, $W$, and the notch width should be 0.015 $W$ or thinner. Based on a span-to-width ratio, S/W, equal to 4, the coefficients, in Equation (2.2) were obtained from Brown and Srawley (1966). The advantage of using a compact tension specimen is that less material is required, but on the other hand, the machining costs are higher and the test setup is more complex compared to the SENB test setup (Figure 2.2).

The presence of pore space, also known as *porosity*, complicates the failure mechanics and makes it a difficult task for precise analysis. The three-dimensional network consisting of cell faces, cell edges, and fluid space in the form of open- and closed-cell foam structure would deform in a way hard to identify. Gibson and Ashby (1988) used honeycombs for the regular cell geometry and data obtained by various groups to analyze the fracture toughness and obtained the expression for fracture toughness, $K^f_{Ic}$, in terms of the fracture strength of the cell walls, $\sigma_{fs}$, and the relative density, $\rho_f/\rho_s$.

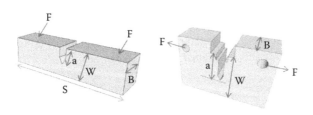

**FIGURE 2.2 (See color insert.)** Typical specimen geometries for SENB and CT tests.

Open-cell foams:

$$K_{Ic}^f = CK_{Ic}\left(\frac{\rho_f}{\rho_s}\right)^{3/2} \tag{2.6}$$

As suggested by Gibson and Ashby (1988), the effects of membrane stress and the fluid (gas for most of the cases) pressure inside a closed-cell foam, although small, should not be ignored in the calculation, hence the term $\varphi$ (the volume fraction of solid contained in the cell edges) into Equation (2.5):

$$K_{IC}^f = C'K_{Ic}\left(\frac{\rho_f}{\rho_s}\right)^{3/2}\phi(1+1.6\phi)^{1/2} \tag{2.7}$$

Where $K_{Ic}$ is the fracture toughness of the solid material measured using conventional methods. C contains all the constants of proportionality and was determined as 0.65 experimentally for both cases.

## 2.2   GENERAL TOUGHENING MECHANISMS

Studies carried out over the years gave detailed theoretical understanding on materials toughening with rubber toughened thermoset systems being the most extensively studied. These studies proposed and provided detailed toughening mechanisms through simulation and actual experimental works to explain the improved toughness achieved by the incorporation of second phase particles (Faber and Evans, 1983; Kawaguchi and Pearson, 2004; Lange, 1970; Pearson and Yee, 1993; Rose, 1987; Shum and Hutchinson, 1990; Zhao and Hoa, 2007). Some important mechanisms are summarized here.

### 2.2.1   PARTICLE BRIDGING

Sigl et al. (1988) proposed that a second phase particle lying on the path of a propagating crack would most likely lead to two phenomena: (1) crack bridging of the particle, especially ductile particles such as rubber, will be stretched and apply a resistive force, which is able to close or suppress crack growth through compression. (2) While bridging the crack, the plastically deformed particles would absorb an additional amount of fracture energy providing a crack shielding effect. Sigl et al. (1988) identified particle bridging as the main contributing mechanism for toughening, whereas the shielding resulting from the yielded particles are negligible. The magnitude of the work done by the particles is dependent on the extent of stretching and the effective distance from the crack (Ahmad, Ashby, and Beaumont, 1986). The compressive force increased the fracture energy until the point of failure; hence large particles are favored in this mechanism (Figure 2.3).

### 2.2.2   CRACK PINNING

Many studies on the fracture toughness of particulate composites consider the crack-bowing mechanism as the main toughening mechanism. Lange (1970) first

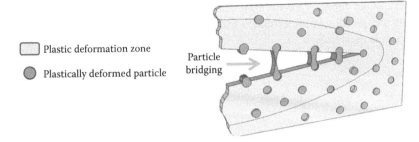

FIGURE 2.3 (See color insert.) Schematic diagram of particle bridging-toughening mechanism. The bridging particles tend to close the crack while surrounding particles in the plastic deformation zone absorb fracture energy through plastic deformation.

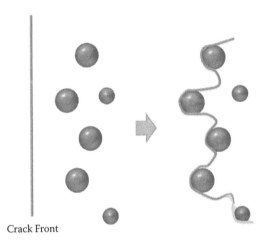

FIGURE 2.4 Schematic representation of the crack pinning-toughening mechanism. The propagating crack front bows out and pins at the particle sites.

proposed the crack-front-bowing mechanism, which proclaimed the resistance of rigid particles to crack propagation. Because of the impenetrability of the particles, the primary crack was perturbed and had to bend between particles into a nonlinear crack front and eventually pinned at the filler sites. The stress intensity of the matrix is reduced by the crack front bowing while the reinforcing phase produces an increase in the stress intensity. The bowed secondary crack front has more elastic energy stored than the straight unbowed crack front. Therefore, more energy is needed for a crack to propagate. When the stress intensity increases until fracture of the reinforcing phase, the crack continues to advance (Figure 2.4).

The *tails*, which are observed behind the particles on the fracture surface, are a characteristic structure created by the mismatch between two planes of crack propagation. The tails are often regarded to be the evidence for the action of the crack-front-bowing mechanism.

Rose (1987) used modified epoxy systems to refine the concept brought up by Lange (1970) and Faber and Evans (1983) to be in better agreement with experimental observations, as represented by Equation (2.5). The final model shows that crack bridging and pinning are interrelated and it is difficult to examine the contribution from a single mechanism.

$$\frac{K_c}{K_0} = \frac{\left[\left(2s/\lambda\right) + \left(2r/\lambda\right)\left(K_L/K_o\right)^2\right]^{1/2}}{F_{kl}} \tag{2.8}$$

Where $K_c$ is the fracture toughness of the modified epoxy, $K_o$ is the fracture toughness of unmodified epoxy, $2s$ is the surface–surface particle spacing, $2r$ is the particle diameter, $K_L$ is a limiting stress factor that specifies the failure of the tailing end of the reinforced zone, and $F_l$ is an interpolating function constructed to reproduce the correct asymptotic expansion for soft and hard spring. It was found that small glass particles comply well with this mechanism and that fracture toughness predictions using this toughening model are in good agreement with experimental results for epoxy toughened with glass spheres.

### 2.2.3 CRACK PATH DEFLECTION

Cracks are usually subdivided into mode I, mode II, and mode III, based on the crack surface displacement. Mode I is the tensile mode and the most commonly encountered mode. The crack tip is subject to displacements perpendicular to the crack plane. In mode II or the sliding mode, the crack faces move relative to each other in the crack plane. Mode III is also referred to as the *tearing mode*, where the shear stress is acting parallel to the plane of the crack and parallel to the crack front (Figure 2.5).

Cracks in the polymer resin advance in a direction with the lowest toughness. When subjected to stress intensity lower than the corresponding planar crack, under the presence of second phase particles, the crack propagation deviates from its main plane into a nonplanar crack. The enhanced toughness comes from the surface energy associated with the newly initiated nonplanar crack surface. Such deviation also reduces the mode I characteristic of the crack opening and induces mode II fracture locally. It is known that most polymeric materials are more resistant to fracture under shear (mode II) loading than tensile (mode I) loading. Therefore, by changing the crack path from mode I to a mixed-mode fracture path, the fracture

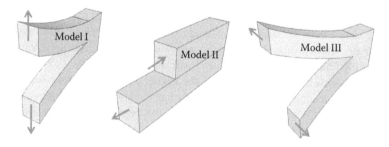

**FIGURE 2.5 (See color insert.)**   Schematic representation for the crack modes.

energy absorbed by this process can be significantly increased. Faber and Evans (1983) proposed models to quantitatively predict the fracture toughness increase due to crack deflection using the fracture mechanics approach. The models demonstrated the mechanism to be size independent and rods of high aspect ratio to be most effective for deflecting cracks.

### 2.2.4  PARTICLE-INDUCED SHEAR YIELDING

This mechanism was identified as the main source of energy dissipation in rubber toughened epoxy systems (Kinloch et al., 1983). Loading at the crack tip led to the occurrence of two interacting processes in a rubber-toughened epoxy system; shear yielding deformation in the matrix and void formation of the rubber particles in response to the triaxial stress ahead of the crack tip. The first process was suggested as the major toughening mechanism, especially when the rubber cavitation increases the stress concentration and reduces constrain on the matrix at its vicinity, which in turn accelerates shear yielding. The material's ability to resist further fracture was improved when the extensive shear yielding suppresses fracture by blunting the crack tip.

### 2.2.5  MICROCRACKING MECHANISM

During processing, especially for the addition of a second phase particle, microcracks are introduced into the system. The nucleation of a microcrack is accompanied by the release of residual stress, which leads to stress redistribution around the main crack tip. Hutchinson relates this phenomenon to the plastic deformation of crack tips in a metal or to the dilatational transformation at the crack tips in a ceramic (Shum and Hutchinson, 1990). The generation of new crack surfaces absorbs fracture energy, which in turn suppresses further crack propagation. Rubber is a typical ductile particle commonly used to toughen polymer matrix by the crack shielding effect resulting from microcrack formation.

Crazing usually occurs when a number of microvoids nucleate at high stress concentration points (i.e., small cracks, air bubbles, voids). The microvoids tend to develop along the plane perpendicular to the maximum principal stress direction, and instead of coalescing to form a crack, they remain separated by fibrils of plastically deformed material, forming a craze.

Characterization, by means of SEM, revealed different types of microcracks on the fracture surface which complicate the mechanism, especially when the second particulate phase is involved in the fracture event: (1) microcracks in the matrix, (2) microcracks in the particle, and (3) along the particle/matrix (debonding) (Lee and Yee, 2001a,b). The fracture toughness can be further influenced by various toughening mechanisms, such as crack deflection, crack bowing, particle debonding (Anderson, Leevers, and Rawlings, 2003; Zhang and Ma, 2010), interparticle spacing theory (Lee and Yee, 2000), and step structure mechanisms (Lee and Yee, 2001a,b).

Figure 2.6 illustrates crack bowing, crack deflection, debonding, and step structure mechanisms, which may be operative to a system. Figure 2.6a shows the occurrence of the crack-bowing mechanism when the propagating crack interacts with fillers

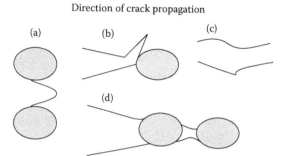

**FIGURE 2.6**   Schematic of proposed fracture mechanism of a composite system: (a) crack-bowing mechanism; (b) crack deflection mechanism; (c) step structure mechanism, and (d) debonding mechanism.

of sufficient strength to resist crack penetration causing the crack front to bow out. Figure 2.6b shows the crack deflection mechanism, which was discussed previously.

During crack bowing, there will be a point at which the crack front breaks away from the rigid particle. At this point, the secondary crack fronts will come together and form a characteristic step structure as both secondary crack fronts propagate at a different crack plane. The step structures, shown in Figure 2.6c, are considered to be the evidence of the existence of the nonlinear and/or nonplanar crack front. When two secondary crack fronts separated by a particle meet with each other, the different kinds of step structures would be formed as both secondary crack fronts propagate at different crack planes. Several kinds of step structures were observed at lower magnification such as *lance*, *river*, and *hackles* (Lee and Yee, 2001a,b). The step structures are thought to be formed due to the mixed mode of fracture in a constrained crack propagation situation.

The debonding mechanism shown in Figure 2.6d is a complex process. After debonding, the modulus of the material is effectively lowered in the process zone around the crack tip and consequently stress intensity is reduced. Therefore, this acts as a crack-speed decreasing mechanism, which is able to absorb significant amounts of fracture energy. In this case, small particles act as better tougheners than large ones.

The debonding mechanism is affected by many factors, such as the size of the particle and the interfacial adhesion. A simple equation was proposed to predict the relationship between the critical stress ($\sigma_i$) and the radius of the particles (Pukanszky et al., 1994):

$$\sigma_i = \frac{-\sigma_t + 2G_a G_0}{|\sigma_t| kr} \tag{2.9}$$

where $G_a$ is the bond fracture energy per unit of bonded surface, $k$ is a constant, $r$ is radius of the microsphere, $G_0$ is shear modulus of the matrix, and $\sigma_t$ is thermal stress. Although the equation cannot precisely predict $\sigma_i$, it indicates that if the size of a microsphere decreases, the critical stress, which is necessary to cause debonding, increases, resulting in less debonding.

The strength of the adhesive bond between the particle and the matrix also determines the level of stress transfer across the interface under loading. In order to improve the adhesion, it is possible to use coupling agents, which could create improved chemical bonds between the particle and matrix in order to further enhance the toughening effect under loading. Silanes are often used in glass–polymer systems. The syntactic foam containing silane coating microspheres exhibited 25% higher strength than that containing untreated microspheres under flexural loading (Koopman et al., 2006). Another coupling agent, glutaric dialdehyde, was used in a carbon microspheres/phenolic resin system (Zhang and Ma, 2010). Syntactic foam containing coupling agent-treated hollow carbon microspheres outperform syntactic foams that contain untreated hollow carbon microspheres under compression and flexural loading.

The interparticle spacing theory was first described in 2000 (Lee and Yee, 2000). The interparticle separation between particles will decrease with increasing particle content. As long as the particle content is beyond the complete wetting ability of the resin, interparticle sliding and stress concentration would be introduced. The chances of particle debonding would increase when the system contains a higher volume fraction of particles. Debonding is usually accompanied by premature microcracks. If the crack growth direction is parallel to these microcracks, the subcritical cracks act as precursors and initiate crack propagation.

The above mechanisms were targeted to provide insights for the existing toughening mechanisms proposed for bulk resins and are not necessarily able to provide accurate quantitative measurements for foam structures, nevertheless, they allow reasonable predictions to be made in toughness that one can achieve if one were to toughen the foam structure by developing nanocomposite foams.

## 2.3  TOUGHENING OF POLYMERIC FOAMS

Polymeric composite foam is a ternary material comprised of polymer matrix, gaseous phase defining the voids (cells), and third phase particle filler. Incorporation of a thermosetting matrix, that is, epoxy resin, in syntactic foams results in many useful properties for structural, engineering applications, such as a high modulus and failure strength, low creep, and good performance at elevated temperatures. However, the structure of such relatively brittle material polymers in combination with voids leads to a highly undesirable property of poor resistance to crack initiation and growth. The low resistance to crack initiation and growth limits the number of applications of these materials in areas prone to high stresses and shock loading. In order to increase the number of advanced applications of foam-based composites, it is essential to increase the toughness of these types of foams.

The toughening mechanisms summarized earlier on provide good insight into possible mechanisms responsible for toughness improvement through second phase infusion. The hypotheses can be applied to the ternary system of foam structures where the cell struts show similar behavior as bulk composite materials. It is hard to identify the fracture and toughening mechanism for the complex three-dimensional foam structure consisting of open/closed-cell, cell walls, and struts. Here the effects of microtoughening and nanotoughening are illustrated in detail using epoxy syntactic foams prepared by Wouterson et al. (2004, 2005, 2007a, 2007b). The results of various reinforcements of syntactic foam and their effect on fracture properties will be discussed in light of microstructure and fracture mechanics.

### 2.3.1 MICROTOUGHENING THROUGH CONTROLLED POROSITY

The presence of cells marks the distinctive difference between bulk composite and composite foams. The level of porosity measures the amount of *empty* space within the matrix and varies with foam density. For the case of syntactic foams, different microstructures or levels of porosity can be created through varying the type and amount of microspheres. Details of the microspheres used by Wouterson et al. (2007b) to prepare the epoxy syntactic foams are listed in Table 2.1. In the fracture toughness assessment under quasistatic loading, SENB specimens were loaded in a three-point bend (3PB) geometry. Due to the difference in density between the various types of microspheres, densities of foams with equivalent amounts of microspheres vary. The difference in density makes the comparison of the properties of foams nonrelevant. In order to compare the performance of foams, the specific mechanical/fracture properties are used.

Figure 2.7 shows the specific fracture toughness, $K_{Ic}/\rho$, and specific energy release rate, $G_{Ic}/\rho$, respectively, for various compositions of syntactic foam. The increase in $K_{Ic}/\rho$ suggests the presence of a toughening mechanism. For $K_{Ic}$, the hollow glass microspheres outperform the hollow phenolic microspheres with K46 microspheres resulting in the highest value for $K_{Ic}/\rho$, suggesting that the wall thickness and density of the microspheres affect the fracture toughness of the syntactic foam. The hollow phenolic microspheres outperform the hollow glass microspheres for the specific energy release rate. This is mainly caused by the lower value of the Young's modulus of the BJO-093 samples.

The drop for $K_{Ic}$ and $G_{Ic}$ after 30 vol% of microspheres indicates the change in the type of major toughening mechanism at the transition point. SEM micrographs elucidated the toughening mechanisms in the various microstructures of syntactic foam. Figure 2.8a shows the fracture surface of pristine epoxy resin after being

### TABLE 2.1
#### Properties of Various Types of Microspheres

| Microsphere | Density | Static Pressure | Mean Diameter | Wall Thickness |
|---|---|---|---|---|
| | (g/cm³) | (MPa) | (μm) | (μm) |
| BJO-093 | | | | |
| Phenolic microspheres | 0.25 | 3.44 | 71.5 | 1.84 |
| K15 | | | | |
| Glass microsphere | 0.15 | 2.07 | 70 | 0.70 |
| K46 | | | | |
| Glass microsphere | 0.46 | 41.37 | 43.6 | 1.37 |

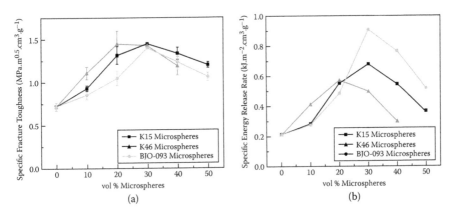

**FIGURE 2.7**    Specific fracture toughness and specific critical energy release rate for various compositions of syntactic foam.

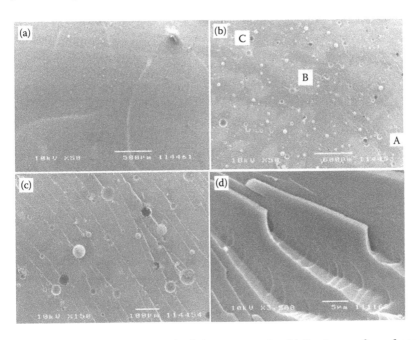

**FIGURE 2.8**    (a) Fracture surface of pristine epoxy resin. (b) Fracture surface of epoxy syntactic foam showing precrack, process, and fast fracture zone. (c) Step structures on the fracture surface of syntactic foam. (d) Step structures on the fracture surface of syntactic foam.

subjected to 3PB. The featureless fracture surface of epoxy resin is indicative of the well-known brittle deformation of epoxy resin. Uniformly oriented white lines on the fracture surface suggest minor shear deformation of the epoxy resin prior to fracture.

The fracture surface of brittle particulate-filled composites can be divided into three areas, namely, precrack, process zone, and fast fracture zone. Tapping a fresh razor blade into a prenotch creates the precrack zone. The sharp crack is required to eliminate any deformation induced by the introduction of the prenotch. The process zone is the area where the crack undergoes stable subcritical crack growth. The process zone is considered as the most important area of the fracture surface in terms of contribution to the overall fracture toughness.

An example of a fracture surface of a syntactic foam containing 5 vol% K46 glass microspheres showing the precrack zone (A), process zone (B), and fast-fracture zone (C), is shown in Figure 2.8b. The gentle roughness of the fracture surface, as shown in Figure 2.8c, is caused by the so-called step structures, which are shown at higher magnification as diagonal lines in Figure 2.8d. In particular, the *tail* structure behind the hollow glass microspheres can be clearly identified, evidence for the action of the crack-front-bowing mechanism. The mismatch between the two crack planes is also clearly visible in Figure 2.8d.

In addition to the presence of step structures, debonding of microspheres is observed on the fracture surface of syntactic foam (see Figure 2.9). Debonding can be identified by the absence of the microsphere or by a gap in between the microsphere and the matrix. The presence of debonding of microspheres from the polymer matrix suggests the weak interfacial adhesion at the microsphere–matrix interface. Several researchers have suggested that the debonding of glass beads is one of the major toughening mechanisms that triggers matrix plastic deformation in glass bead reinforced polymers. The theory of debonding and matrix plastic deformation shows similarity with the presence of cavitations of rubber particles and matrix shear yielding in rubber toughening of polymers. Debonding involves energy dissipation and will thus impede the crack growth process.

For some syntactic foams with higher volume fractions of microspheres, crushed microspheres become evident. An example of a crushed microsphere is given in Figure 2.9b. Once the microspheres crush, energy is dissipated in shattering the microsphere, resulting in relief of triaxial tension and crack path deflection, attributing to the overall fracture toughness of the syntactic foam.

Figure 2.10 shows the fracture surface of syntactic foam containing low and high volume fractions of microspheres, respectively. It is clearly shown that step structures prevail for the microstructures containing low volume fractions of microspheres. Step structures are absent in the microstructures of syntactic foam containing 50 vol% hollow microspheres. Instead, the dominant fracture mechanism that

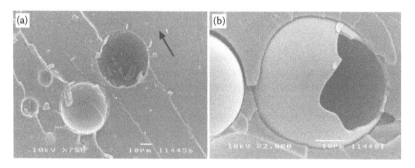

**FIGURE 2.9**    (a) Debonding of microspheres from epoxy matrix. (b) A debonded and broken microsphere.

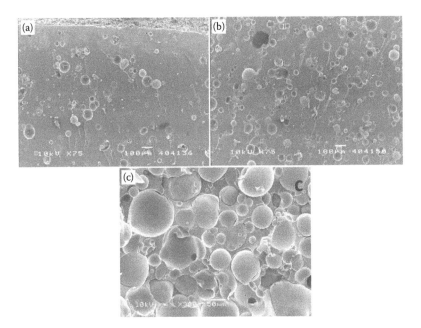

**FIGURE 2.10** Fracture surface of syntactic foam containing (a) 10 vol%, (b) 20 vol%, and (c) 50 vol% of microspheres.

is observed is debonding of microspheres. According to Lee and Yee (2001a,b), the increase in particle content will decrease the interparticle separation. The increase of particle content beyond the complete wetting ability of the matrix introduces inter-sphere sliding. Higher volume fractions of particles would allow more debonding from the matrix. If the direction of these cracks is parallel to the crack growth direction, the subcritical cracks act as precursors, and facilitate crack propagation.

Understanding of the change in a toughening mechanism between low and high volume content of microspheres explains why there is a maximum value in the specific fracture toughness versus microsphere content curve. What is left is an explanation for the increase in the fracture toughness of syntactic foams containing 0–30 vol% microspheres. Comparing Figure 2.10a,b and Figure 2.10c, it is clear that similar toughening mechanisms can be observed for syntactic foam with higher volume fractions of hollow microspheres. The distinctive features observed are that the length of the step structure and the intensity of the tail structures have now increased due to the larger numerical density of microspheres causing the fracture toughness to increase with an increase in volume fraction of the microspheres.

## 2.3.2 MICROTOUGHENING THROUGH SHORT FIBER INCLUSION

It is well understood that short fibers are an effective means to toughen polymers. However, for polymeric foams, short fiber reinforcement is relatively unknown. 1, 2, and 3 wt% of short carbon fiber (SCF) of mean fiber lengths 3.11, 4.50, and 10.05 mm were added to the epoxy resin to study the effect of the fibers on the

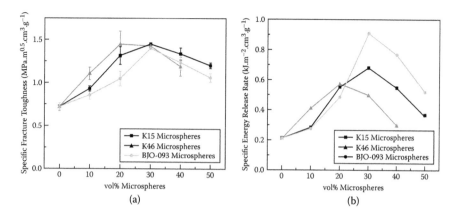

**FIGURE 2.11**   General microstructure of short carbon fiber-reinforced syntactic foam.

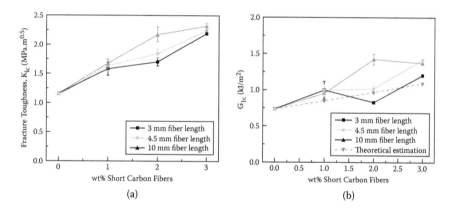

**FIGURE 2.12**   Fracture toughness ($K_{IC}$) and critical energy release rate ($G_{IC}$) of short fiber-reinforced syntactic foam.

fracture properties. Figure 2.11 shows typical scanning electron micrographs of short fiber-reinforced syntactic foam (SFRSF).

Figure 2.12 shows the results of the effect of fiber length and fiber weight fraction on the fracture toughness and critical energy release rate of SFRSF determined by Equation (2.1) to Equation (2.3). The increase in the plane strain fracture toughness, $K_{Ic}$, with increasing fiber content strongly suggests the synergistic benefit of fiber reinforcement. The effect of the fiber lengths on the fracture properties appears to be less pronounced. This observation is in line with the work reported by Chiang (2000), who highlighted that the fracture toughness reaches a limit if the fiber reaches the critical fiber length.

A SEM fractograph of a SENB specimen is shown in Figure 2.13. The black arrow denotes the direction of the crack propagation. Various toughening mechanisms, including fiber pull out, fiber breakage, step structures, debonding of microspheres, and fractured microspheres are identified. An image of the sharp crack and crack

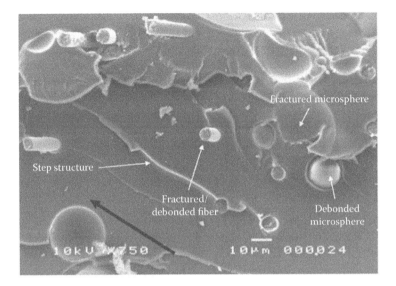

**FIGURE 2.13**    Fracture surface of short carbon fiber-reinforced syntactic foam.

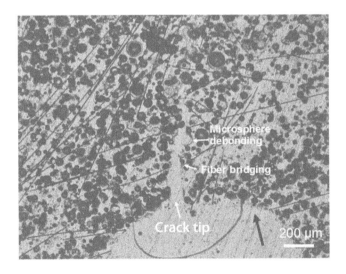

**FIGURE 2.14**    Optical micrograph of crack tip area in short fiber-reinforced syntactic foam.

initiation region obtained with a transmission optical microscope (TOM) is shown in Figure 2.14. The image clearly shows the presence of fiber bridging and debonding of microspheres at the crack interface. Despite observing multiple specimens containing different amounts of reinforcing fibers under the optical microscope, no microshear banding was observed under polarized light. Microshear banding has been reported by Lee and Yee (2001a,b) as an important toughening mechanism in glass sphere reinforced polymers but could be absent in brittle matrices.

The presence of multiple toughening mechanisms on the fracture surface of SFRSF SENB specimens requires an in-depth analysis to determine the contribution of each toughening mechanism to the overall toughness of the composite. The presence of fiber breakage, interface debonding, and pull out suggests the presence of fiber bridging during the fracture process. As the crack front approaches, it is halted by the fiber and shows a tendency to circumvent the crack face in the transverse direction, causing interfacial debonding between the fiber and the matrix. The contribution of the interfacial debonding toughness, $R_d$, to the overall SFRSF toughness can be estimated by using the following equation (Kim and Mai, 1998):

$$R_d = \frac{V_f \sigma_f^2 l_d}{2E_f} \qquad (2.10)$$

where $V_f$ is the volume fraction of SCF, $l_d$ the debonding length, and $E_f$ the Young's modulus of the SCF. Values for $R_d$ obtained for SFRSF containing various amounts of SCF are given in Table 2.2. The poor interfacial debonding is confirmed by the matrix-free fiber surface observed in Figure 2.13, indicating a poor interface between the matrix and the fibers.

After the interfacial debonding, fibers can either debond from the matrix or fracture in the crack plane, depending on the fiber aspect ratio. According to Hull and Clyne (1996), debonding of fibers occurs if the fiber aspect ratio, s ( = l/d), is below a critical value. The critical value, $\sigma$crit, can be calculated using the following equation (Hull and Clyne, 1996):

$$\sigma_{crit} = \frac{\sigma_f}{2\tau_m} \qquad (2.11)$$

where $\tau_m$ is assumed to be the matrix shear strength. For the SCF dispersed in the SFRSF, $\sigma_{crit}$ is estimated to be 28.2. Based on a fiber diameter of 7 µm, the fiber aspect ratios for the three different fiber lengths of 3, 4.5, and 10 mm, are estimated to be 429, 714, and 1428, respectively. It is obvious that $\sigma_{crit}$ is only a fraction of s for all the three fiber lengths, indicating that the SCF in SFRSF will fracture in the crack plane instead of fully debonding from the matrix. Fracture of the fibers will prevail since there will always be embedded lengths on either side of the crack plane which are long enough for the stress in the fiber to build up sufficiently to break the fiber. According to Hull and Clyne (1996), the actual fiber breaking makes little or no direct contribution to the overall toughness. Typical fracture energies for carbon fibers are only a few tens of kJ/m².

Though the actual breaking of the fiber is expected to have a minor effect on the overall fracture toughness of the fiber-reinforced composite, it is believed that the redistribution of the strain energy that occurs when a fiber breaks might have more impact on the overall toughness. According to Kim and Mai (1998), the toughness due to the redistribution of the strain energy stored in the fiber prior to fracture, $R_r$, can be calculated with the following equation:

$$R_r = \frac{V_f \sigma_f^2 l_{crit}}{3E_f} \qquad (2.12)$$

**TABLE 2.2**

**Conjectural Estimates of Toughening Contributions from Interfacial Debonding, $R_d$, Strain Redistribution Energy, $R_r$, Fiber Pull Out, $R_{po}$, and Formation of New Surfaces, $R_s$, to the Total Toughness of Short Fiber-Reinforced Syntactic Foam**

| SCF Weight Fraction (wt%) | $R_d$ (kJ/m²) | $R_r$ (kJ/m²) | $R_{po}$ (kJ/m²) | $R_s$ (kJ/m²) | $R_t$ (kJ/m²) |
|---|---|---|---|---|---|
| 0 | NA | NA | NA | 0.736 (100%) | 0.736 |
| 1 | 0.0006 (~0.07%) | 0.0082 (~0.96%) | 0.0122 (~1.43%) | 0.8334 (~97.54%) | 0.8544 |
| 2 | 0.0012 (~0.12%) | 0.0164 (~1.68%) | 0.0245 (~2.52%) | 0.9319 (~95.68%) | 0.9740 |
| 3 | 0.0019 (~0.17%) | 0.0247 (~2.26%) | 0.0370 (~3.38%) | 1.0313 (~94.19%) | 1.0949 |

*Note:* The relative % from each contribution is indicated by the value in parentheses.

Table 2.2 shows the conjectural estimates for $R_r$ and several other toughness contributions. The relative contribution from each mechanism is shown by bracketed values. Clearly, the contribution of $R_r$ to the overall toughness is relatively low, which is mainly caused by the low volume fraction of SCF in SFRSF.

Despite the fact that fibers show the preference to fracture in the crack plane, debonding and pull out of fibers can be observed on the fracture surface of SENB specimens (see Figure 2.13). However, the contribution of debonding to the overall toughness is minimal as the debond length, as estimated from SEM images, is only in the order of tens of micrometers. It should be noted that the amount of debonding is related to the fiber orientation. Fibers oriented perpendicular to the crack plane show the smallest debonding length. The contribution of the fiber pull-out toughness, $R_{po}$, to the toughness of SFRSF can be estimated as follows (Kim and Mai, 1998):

$$R_{po} = \frac{V_f \tau_{fr} l_{crit}^2}{6d} \frac{l_{crit}}{l} \tag{2.13}$$

where $\tau_{fr}$ is the frictional shear stress. Please note that Equation (2.13) is for unidirectional fibers having a length greater than the critical fiber length. Further, it is clear from Figure 2.13 that localized matrix yielding occurs near fiber ends. In addition, step structures can be observed behind many of the fibers. The quantity and intensity of the step structures observed on the fracture surface of SFRSF seems to be more significant compared to pristine syntactic foam. As discussed before, step structures are the result of the crack-front-bowing mechanism as proposed by Lange (1970). Many studies on the fracture toughness of particulate composites consider the crack-bowing mechanism as the main toughening mechanism. The main aspect responsible for the absorption of the energy is the formation of new surfaces. The contribution of the toughness, due to the formation of new surfaces, $R_s$, to the overall toughness of fiber-reinforced composites, was discussed by Marston et al. (1974) and Atkins (1975). Rs is regarded as the sum of the specific energies absorbed in creating new surfaces in fiber, matrix, and at the interface. For SFRSF, additional surfaces are created in microspheres. To account for the additional microsphere surfaces, the equation given by Atkins (1975) has been modified by replacing the matrix toughness with the toughness of pristine syntactic foam containing 30 vol% phenolic microspheres, $R_{sf}$. $R_s$ can then be calculated with the following equation:

$$R_s = V_f R_f + (1 - V_f) R_{sf} + \frac{V_f}{l_{crit}} R_{ic} \tag{2.14}$$

By assuming that the fiber toughness, $R_f$, is negligible and the interfacial fracture toughness, $R_{ic}$, is equal to the toughness of pristine syntactic foam, $R_{sf}$, Equation (2.14) can be rewritten to the following:

$$R_s = \left[ V_f \left( \frac{l_{crit}}{d} - 1 \right) + 1 \right] R_{sf} \tag{2.15}$$

The estimated values for $R_s$ can be found in Table 2.2. It is obvious that the specific energy required for the formation of new surfaces is significantly higher compared to any of the other toughening mechanisms discussed and can, thus, be considered as the main toughening mechanism in SFRSF.

The total toughness of SFRSF, $R_t$, is now estimated by combining the contributions of each individual mechanism. $R_t = R_d + R_r + R_{po} + R_s$.

Figure 2.12b reflects the result of $R_t$ for various weight fractions of SCF. It is clear that the conjectural estimate of $R_t$, as discussed, represents the trend and values for the experimentally measured critical energy release rate versus weight fraction of SCF quite closely.

### 2.3.3 NANOTOUGHENING OF POLYMERIC FOAM

Experimental work on nanocomposites generally demonstrated new or improved properties compared to the micro- and macrocomposite counterparts. Nanotechnology has proven to be an effective tool in increasing the fracture properties of materials (Kinloch and Taylor, 2006). Dispersion of particles in a polymer matrix usually results in the presence of stress concentrations, in particular when the moduli and Poisson's ratio are different from the matrix. The particles act as microcrack initiation points or may debond, resulting in increased toughness. As the size of the particles moves toward the nanometer scale, one may wonder whether finer particles eventually will lead to even better toughening effects.

The inclusion of nanoclay in polymeric materials has been proven to elucidate remarkable tensile, fracture, and thermal properties enhancements (Lan and Pinnavaia, 1994; LeBaron, Wang, and Pinnavaia, 1999), attributed to the high aspect ratio of the silicate nanolayers. The largest improvement in properties are observed for those nanocomposites where the polymer is able to penetrate and expand the gallery in between two clay layers. This is referred to as *exfoliation*. Often, the polymer only slightly expands the gallery, which is called *intercalation*. Though both intercalated and exfoliated nanoclay layers are able to improve the properties of the nanocomposites, the highest improvements have been observed for exfoliated nanoclay layers.

Despite being proven to be an effective reinforcement for polymers, nanoclay-reinforced syntactic foam (NCRSF) has only been explored in recent years (John, Nair, and Ninan, 2010; Maharsia, Gupta, and Jerro, 2006; Peter and Woldesenbet, 2008). It would be interesting to evaluate a possible synergistic behavior in hybrid syntactic foams comprised of a matrix, hollow microspheres, and nanoclay particles.

Similar to the studies performed on the SFRSF, the amount of microspheres, for each nanoclay-reinforced syntactic foam, was fixed at 30 vol%. The nanoclay was surface-modified montmorillonite mineral. Figure 2.15 shows the general structure of NCRSF.

Clay intercalations were observed inside the epoxy matrix as shown in Figure 2.16. Figure 2.16b shows increased exfoliation of the nanoclay when 1.0 wt% of nanoclay

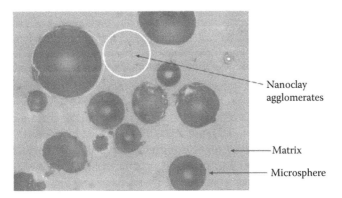

**FIGURE 2.15**  Reflective optical micrograph showing the general structure of nanoclay-reinforced syntactic foam.

**FIGURE 2.16**  TEM image of (a) 0.5 and (b) 1.0 wt% nanoclay-reinforced syntactic foam at higher magnification.

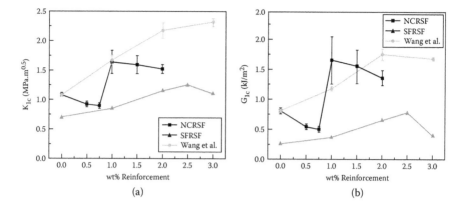

**FIGURE 2.17** Fracture toughness and critical energy release rate of nanoclay-reinforced syntactic foam.

was dispersed. At lower magnification, nanoclay agglomeration was also observed with the degree of agglomeration increased with increasing nanoclay loadings.

Figure 2.17 shows the fracture toughness, $K_{Ic}$, and critical energy release rate, $G_{Ic}$, of NCRSF, respectively. For NCRSF, containing 1–2 wt% nanoclay particles, an increase in $K_{Ic}$ is observed compared to pristine syntactic foam. Agglomeration of the nanoclay layers at higher loading content caused the $K_{Ic}$ value to drop.

The critical energy release rate, $G_{Ic}$, for syntactic foam containing various amounts of nanoclay is shown in Figure 2.17b. The trend in $G_{Ic}$ versus wt% nanoclay is similar to the trend observed for $K_{Ic}$. Introduction of 0.75 wt% causes a reduction of 41% for $G_{Ic}$. Compared to pristine syntactic foam, $G_{Ic}$ shows a significant increase of 104% for specimens containing 1 wt% nanoclay. The sharp increase at 1 wt% nanoclay suggests that 1 wt% of nanoclay acts like a threshold; the amount of nanoclay required to induce toughening in NCRSF. The increase in both $K_{Ic}$ and $G_{Ic}$ for syntactic foam containing 1 wt% nanoclay indicates the excellent toughening potential of nanoclay in syntactic foams.

For the syntactic foams mentioned above, in general, for similar weight fractions, microtoughening is more effective than nanotoughening. A comparison, see Figure 2.17, was made against the fracture data for highly exfoliated nanoclay-toughened epoxy reported by Wang et al. (2005). It is obvious that NCRSF features a superior fracture behavior compared to nanoclay-toughened epoxy. The result suggests that an exfoliated nanoclay system might not result in the optimal fracture behavior. Agglomeration of nanoclay layers is preferred as the agglomerates are able to divert the path of crack growth and induce the formation of microcracking, increasing the fracture toughness. The improved fracture behavior of NCRSF over nanoclay-toughened epoxy suggests a synergistic behavior between the hollow microspheres and the nanoclay layers.

The fracture surfaces of the NCRSF specimens containing 0.75 wt% and 1 wt% nanoclay were examined using SEM (see Figure 2.18) to elucidate the toughening mechanisms responsible for the fracture behavior of nanoclay-reinforced syntactic foams. The three characteristic fracture regions, namely precrack (A), process or crack initiation zone (B), and the fast fracture region (C), are clearly visible in

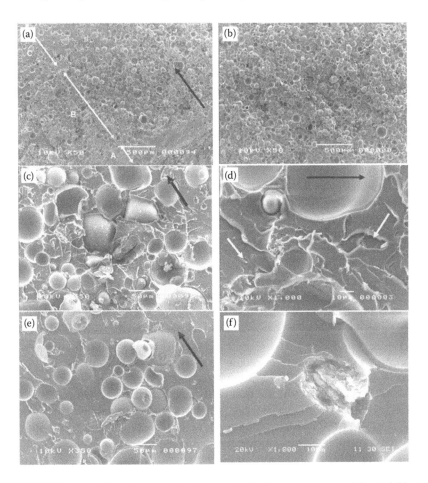

**FIGURE 2.18** Overall fracture surface of a syntactic foam reinforced with: (a) 0.75 wt% nanoclay. (b) Crack process zone of syntactic foam reinforced with 0.75 wt% nanoclay. (c) Fast fracture zone of a syntactic foam reinforced with 0.75 wt% nanoclay. (d) Crack initiation zone of a syntactic foam reinforced with 1.0 wt% nanoclay (low magnification). (e) Crack initiation zone of a syntactic foam reinforced with 1.0 wt% nanoclay (high magnification). (f) Step structures generated by agglomerated nanoclay.

Figure 2.18a. The precrack region is produced by tapping of a fresh razor blade into the prenotch. The process zone is of interest as it harbors the toughening mechanisms responsible for the material's fracture toughness. The amount of fracture toughness is often reflected in the surface roughness and size of the process zone. The fracture surface of the process zone of specimens containing 0.75 wt% nanoclay, as seen in Figure 2.18b, is rough whereas the fracture surface of the fast fracture region, see Figure 2.18c, is relatively smooth. A smooth surface is associated with brittle behavior and does not contribute significantly to the overall composite toughness. According to Zerda and Lesser (2001), the rough surface exhibits evidence for crack branching along the path length. As the clay content increases, the distance between regions of intercalated and agglomerated clay is reduced. The reduced distance causes the crack to follow a more tortuous path, either around or between regions of high clay content. Increased clay content leads to increased crack path deflection, causing increased surface roughness. Indeed, the fracture surface of the process zone of specimens containing 0.75 wt%, as shown in Figure 2.18a, is less rough compared to the fracture surface of specimens containing 1.0 wt% nanoclay (see Figure 2.18d).

The size of the process zone also plays a significant role in the contribution to the fracture toughness. The larger the process zone, the more toughening events were able to take place, the higher the fracture toughness of the composite. The fracture toughness of specimens containing 1.0 wt% nanoclay is higher than specimens containing 0.75 wt% nanoclay.

In the process zones many potential toughening mechanisms are observed, that is, step structures, matrix deformation, microcracks, clusters of nanoclay particles, and fractured and debonded microspheres. In addition to the typical step structures observed behind the microspheres, step structures also occur behind nanoclay agglomerates. Nanoclay agglomerates tend to be in the order of microns, acting more as a micron particle. For higher nanoclay content, the size of the agglomerates tends to grow. If the size of the agglomerate is in the order of microns, the agglomerate is able to resist and deflect the crack front. This behavior is clearly visible in Figure 2.18f, where significant step structures are visible behind the nanoclay agglomerates. Interestingly, the step structures behind the nanoclay particles seem to be interacting with the surrounding microspheres. This kind of interaction between microspheres and nanoclay layers could be a potential evidence of the synergistic behavior between nanoclay and microspheres. Furthermore, the size of the nanoclay cluster affects the size of the resulting step structures. The larger the size of the nanoclay clusters, the larger the step structures, and thus the higher the fracture energy. This could be a possible explanation for the higher toughness values for syntactic foam containing ≥ 1wt% of nanoclay particles.

In addition to step structures, multiple microcracks are also observed in Figure 2.18e (indicated by white arrows). Observations of these microcracks are similar to those reported by Wang et al. (2005). According to Wang et al. (2005), the presence of the microcracks implies that the clay layers act as stress concentrators, promoting the formation of a large number of microcracks upon the loading of the sample, causing an increase in the size of the process zone.

Optical microscopy (OM) was performed to study the subsurface damage in the crack process zone. The study allows for evaluation of the possible presence of microshear banding as reported by Lee and Yee (2000). Microshear banding has been reported as an important toughening mechanism in glass sphere reinforced polymers. Despite observing multiple specimens containing different amounts of nanoclay under the optical microscope, no microshear banding can be seen under polarized light. Similar findings are reported by Wang et al. (2005) for epoxy/clay nanocomposites. The absence of microshear banding could be attributed to the relatively low content of nanoclay layers in combination with their random orientation. Further, it is suggested by the author that the system might behave in an elastic manner causing *healing* of the crack, leaving no traces of microshear banding.

Transmission optical microscopy revealed another toughening mechanism (see Figure 2.19). The image, taken under crossed polarized light, clearly shows

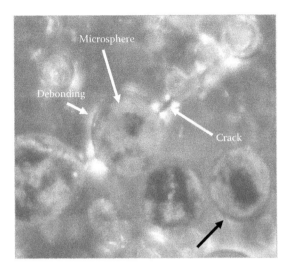

**FIGURE 2.19** Transmission optical micrograph of subsurface damage near the crack initiation area in nanoclay-reinforced syntactic foam.

the presence of diffuse matrix shear yielded regions due to the debonding of the microspheres. Diffuse shear yielding is always found around the debonded matrix. Once a microsphere debonds from the matrix, a free surface will be generated, which is more vulnerable to plastic shear deformation.

## 2.4 CONCLUSIONS

Having discussed microtoughening and nanotoughening in detail through syntactic foams with inclusion of a second phase particle, the following should be considered when analyzing the toughening mechanisms of polymeric foams consisting of a second particle phase:

- Microtoughening of syntactic foam through changing the level of porosity by altering the type of microspheres has proven to be effective. For lower filler content an increase in the specific fracture toughness was observed. The increase reached a maximum after which a decrease in the specific fracture toughness was seen. The change in behavior was attributed to a change in the dominant toughening mechanisms from filler stiffening, crack-front bowing to excessive debonding. This observation can be extended to particle-toughened systems by replacing microspheres to a second phase particle.
- Microtoughening through the inclusion of short carbon fiber is an effective method to increase the fracture toughness of syntactic foams. Fractured and debonded fibers, leading to increased matrix plasticity, step structures, and microsphere debonding are believed to be responsible for the increase in fracture properties.
- From a conjectural estimate of the total fracture toughness, the generation of new surfaces was the highest contributor to the overall toughness.

- Nanotoughening is an effective method to increase the fracture toughness of polymeric foams. Additionally, the nanoclay layers introduced additional crack deflection and microcracking. The intensity of the crack deflection increased with the size of the nanoclay agglomerates.

In principle, the combination of various toughening mechanisms instead of an individual mechanism is involved in a system and results in the failure of specimen. For example, a debonding mechanism would be dominant as long as the increase in interfacial force between filler particles and matrix could result in the improvement in mechanical properties. Crack bowing and deflection mechanisms could be accompanied though the dominant mechanism is debonding. The change of major toughening mechanism could influence the mechanical properties. Therefore, it is important for the scientists and engineers to clarify the toughening mechanisms and apply them to guide future design for composite materials.

## REFERENCES

Ahmad, Z. B., M. F. Ashby, and P. W. R. Beaumont, The contribution of particle-stretching to the fracture toughness of rubber modified polymers. *Scripta Metallurgica*, 1986. 20(6): 843–848.

Anderson, K. D. L., P. Leevers, R. D. Rawlings, ed., *Materials Science for Engineers*, 5th ed. 2003. Boca Raton, FL: CRC Press.

Atkins, A. G., Intermittent bonding for high toughness/high strength composites. *Journal of Materials Science*, 1975. 10(5): 819–823.

Brown, W. F. and J. E. Srawley, Plain strain crack toughness testing of high strength metallic materials. *ASTM Special Technical Publication*, 1966. 410.

Chiang, C. R., Prediction of the fracture toughness of fibrous composites. *Journal of Materials Science*, 2000. 35(12): 3161–3166.

Faber, K. T. and A. G. Evans, Crack deflection processes—I. Theory. *Acta Metallurgica*, 1983. 31(4): 565–576.

Gibson, L. J. and M.F. Ashby, *Cellular Solids Structure & Properties*, 1988, Oxford: Pergamon Press.

Hull, D. and T. W. Clyne, *An Introduction to Composite Material*, 2nd ed., 1996. Cambridge: Cambridge University Press.

John, B., C. P. R. Nair, and K. N. Ninan, Effect of nanoclay on the mechanical, dynamic mechanical and thermal properties of cyanate ester syntactic foams. *Materials Science and Engineering: A*, 2010. 527(21–22): 5435–5443.

Kawaguchi, T. and R. A. Pearson, The moisture effect on the fatigue crack growth of glass particle and fiber reinforced epoxies with strong and weak bonding conditions: Part 2. A microscopic study on toughening mechanism. *Composites Science and Technology*, 2004. 64(13–14): 1991–2007.

Kim, J. K. and Y. W. Mai, *Engineered Interfaces in Fiber Reinforced Composites*, 1998. Amsterdam: Elsevier.

Kinloch, A. J. and A. C. Taylor, The mechanical properties and fracture behaviour of epoxy-inorganic micro- and nano-composites. *Journal of Materials Science*, 2006. 41(11): 3271–3297.

Kinloch, A. J. and R. J. Young, eds. *Fracture Behaviour of Polymers*, 1983. London: Applied Science.

Kinloch, A. J. et al., Deformation and fracture behaviour of a rubber-toughened epoxy: 1. Microstructure and fracture studies. *Polymer*, 1983. 24(10): 1341–1354.

Koopman, M. et al., Microstructural failure modes in three-phase glass syntactic foams. *Journal of Materials Science,* 2006. 41(13): 4009–4014.

Lan, T. and T. J. Pinnavaia, Clay-Reinforced Epoxy Nanocomposites. *Chemistry of Materials,* 1994. 6(12): 2216–2219.

Lange, F. F., The interaction of a crack front with a second-phase dispersion. *Philosophical Magazine,* 1970. 22(179): 983–992.

LeBaron, P. C., Z. Wang, and T. J. Pinnavaia, Polymer-layered silicate nanocomposites: An overview. *Applied Clay Science,* 1999. 15(1–2): 11–29.

Lee, J. and A. F. Yee, Fracture of glass bead/epoxy composites: On micro-mechanical deformations. *Polymer,* 2000. 41(23): 8363–8373.

Lee, J. and A. F. Yee, Inorganic particle toughening I: Micro-mechanical deformations in the fracture of glass bead filled epoxies. *Polymer,* 2001a. 42(2): 577–588.

Lee, J. and A. F. Yee, Inorganic particle toughening II: Toughening mechanisms of glass bead filled epoxies. *Polymer,* 2001b. 42(2): 589–597.

Maharsia, R., N. Gupta, and H. D. Jerro, Investigation of flexural strength properties of rubber and nanoclay reinforced hybrid syntactic foams. *Materials Science and Engineering: A,* 2006. 417(1–2): 249–258.

Marston, T. U., A. G. Atkins, and D. K. Felbeck, Interfacial fracture energy and the toughness of composites. *Journal of Materials Science,* 1974. 9(3): 447–455.

Pearson, R. A. and A. F. Yee, Toughening mechanisms in thermoplastic-modified epoxies: 1. Modification using poly(phenylene oxide). *Polymer,* 1993. 34(17): 3658–3670.

Peter, S. and E. Woldesenbet, Nanoclay syntactic foam composites—High strain rate properties. *Materials Science and Engineering: A,* 2008. 494(1–2): 179–187.

Pukanszky, B. et al., Micromechanical deformations in particulate filled thermoplastics— Volume strain measurements. *Journal of Materials Science,* 1994. 29(9): 2350–2358.

Rose, L. R. F., Toughening due to crack-front interaction with a second-phase dispersion. *Mechanics of Materials,* 1987. 6(1): 11–15.

Rutz, B. H. and J. C. Berg, A review of the feasibility of lightening structural polymeric com- posites with voids without compromising mechanical properties. *Advances in Colloid and Interface Science,* 2010. 160(1–2): 56–75.

Shum, D. K.M. and J. W. Hutchinson, On toughening by microcracks. *Mechanics of Materials,* 1990. 9(2): 83–91.

Sigl, L.S. et al., On the toughness of brittle materials reinforced with a ductile phase. *Acta Metallurgica,* 1988. 36(4): 945–953.

Wang, K. et al., Epoxy nanocomposites with highly exfoliated clay: Mechanical properties and fracture mechanisms. *Macromolecules,* 2005. 38(3): 788–800.

Wouterson, E. M. et al., Fracture and impact toughness of syntactic foam. *Journal of Cellular Plastics,* 2004. 40(2): 145–154.

Wouterson, E. M. et al., Specific properties and fracture toughness of syntactic foam: Effect of foam microstructures. *Composites Science and Technology,* 2005. 65(11–12): 1840–1850.

Wouterson, E. M. et al., Nano-toughening versus micro-toughening of polymer syntactic foams. *Composites Science and Technology,* 2007a. 67(14): 2924–2933.

Wouterson, E. M. et al., Effect of fiber reinforcement on the tensile, fracture and thermal prop- erties of syntactic foam. *Polymer,* 2007b. 48(11): 3183–3191.

Zerda, A. S. and A. J. Lesser, Intercalated clay nanocomposites: Morphology, mechanics, and fracture behavior. *Journal of Polymer Science Part B: Polymer Physics,* 2001. 39(11): 1137–1146.

Zhang, L. Y. and J. Ma, Effect of coupling agent on mechanical properties of hollow carbon microsphere/phenolic resin syntactic foam. *Composites Science and Technology,* 2010. 70(8): 1265–1271.

Zhao, Q. and S. V. Hoa, Toughening Mechanism of Epoxy Resins with Micro/Nano Particles. *Journal of Composite Materials,* 2007. 41(2): 201–219.

# 3 Extrusion of Polypropylene/Clay Nanocomposite Foams

*Kun Wang and Wentao Zhai*

## CONTENTS

## 3.1 INTRODUCTION

Polypropylene (PP) has many desirable and beneficial properties, such as a high tensile modulus, high melting point, low density, excellent chemical resistance, and easy recycling (Vasile and Seymour, 1993). These outstanding properties and a low material cost have made linear PP foams a potential substitute for other thermoplastic foams, such as polystyrene (PS) and polyethylene (PE), in various industrial applications. However, it is challenging to produce linear PP foams with a high expansion ratio due to its weak melt strength and melt elasticity. It is found that the cell walls are not strong enough to bear any extensional force during bubble growth, and thus the bubbles are prone to coalesce and collapse during foam processing. Consequently, the foamed PP products usually have a high open-cell content and nonuniform cell distribution (Burt, 1978; Park and Cheung, 1997; Zhai et al., 2008a), and thus are not good for practical applications.

Many methods have been used to increase the melt strength of PP resins with the aim to improve its foaming behavior, such as long-chain branching (Naguib et al., 2002; Rodríguez-Pérez, 2005; Spitael and Macosko, 2004; Zhai et al., 2008b), cross-linking (Danaei, Sheikh, and Taromi, 2005; Han et al., 2006; Liu et al., 2006; Ruinaard, 2006; Zhai et al., 2008a), and polymer blending and compounding (Doroudiani, Park, and Kortschot, 1998; Naguib et al., 2006; Nam et al., 2002; Okamoto, Nam, Maiti, Kotaka, Hasegawa et al., 2001; Zhai et al., 2008c; Zheng, Lee, and Park, 2010). Several companies have developed commercialized high-melt-strength PP (HMSPP) resins for the production of good PP foams. Due to the price consideration, however, the actual usage of HMSPP in the foaming area is limited. It is difficult to produce PP foams with a high expansion ratio if only linear PP is used. Blending with PE has been used to improve the cell morphology of PP foam (Lee, Tzoganakis, and Park, 1998; Lee, Wang, and Park, 2006; Rachtanapun, Selke, and Matuana, 2003, 2004), but the ability of PE to do so seems limited. Furthermore, blending with HMSPP has been verified as an effective method for producing low-density PP foam, and the addition of 25 wt% HMSPP was found to efficiently improve the cell morphology and expansion ratio of linear PP foams (Reichelt et al., 2003). Currently, the addition of nanofillers has been considered an effective way to improve the foaming behavior of PP (Nam et al., 2002; Okamoto, Nam, Maiti, Kotaka, Nakayama et al., 2001; Zhai et al., 2010; Zhai and Park, 2011; Zhai, 2012; Zheng, Lee, and Park, 2010).

Nanoclay particles have a platelet-like shape and are composed of layers that are 1–2 nm thick. The high aspect ratio and large surface area of nanoclay particles potentially offer good reinforcing efficiency and improved dimensional and thermal stability. The well-dispersed nanoclay is one kind of nucleating agent that reduces the energy barrier of cell nucleation by inducing a local stress variation in polymer/gas solutions ( Leung et al., 2012; Wang et al., 2010; Wong et al., 2011), and can thus enhance cell nucleation and increase cell density significantly. This phenomenon has been observed in various polymer/filler nanocomposite foaming systems, such as PP/clay (Zhai et al., 2010; Zhai and Park, 2011; Zheng, Lee, and Park, 2010), PE/clay (Guo et al., 2007; Lee et al., 2005; Lee, Wang et al., 2007), PS/clay (Han et al., 2003; Shen, Zeng, and Lee, 2005; Zeng et al., 2003), nylon/clay (Yuan, Song, and Turng, 2007; Yuan et al., 2004; Zheng, Lee, and Park, 2006), and polycarbonate (PC)/SiO$_2$ (Zhai et al., 2006). The introduction of well-dispersed nanoparticles can also increase the melt strength of polymer or induce strain hardening of the melt (Koo et al., 2005; Okamoto, Nam, Maiti, Kotaka, Hasegawa et al., 2001; Park et al., 2006), which tends to suppress the occurrence of cell coalescence.

## 3.2 PREPARATION OF POLYPROPYLENE/CLAY NANOCOMPOSITES BY MELT BLENDING

It has been verified that the dispersion behavior of nanoclay strongly affects the foamability of PP/clay nanocomposites. In this section, we briefly introduce the methodology to prepare PP/clay nanocomposites with well-defined clay dispersion.

Two methods have been used to prepare PP nanocomposites, that is, melt blending and *in situ* polymerization. Considering the industry's need for a highly efficient process, melt blending has been considered to be an economical way to fabricate polymer nanocomposites. Consequently, the following discussion focuses mostly on this process.

The point of improving the properties of polymer/clay nanocomposites is how to exfoliate and well disperse individual platelets with high aspect ratios, that is, over 200, throughout the polymer matrix (Ray and Okamoto, 2003). Currently, an exfoliation mechanism for nanoclay had been proposed for the melt-blending method (Dennis et al., 2001). It was shown that as the nanoclay particles were melt blended into a polymer, the clay particles were first fractured by the shear in the extruder. The polymer chains were then diffused into the clay interlayer because of a physical or a chemical affinity between the polymer and the nanoclay surface. Finally, the chains pushed the end of the platelets apart, and an onion-like delaminating process further continued to disperse the platelets into the polymer matrix.

The preparation of exfoliated nanocomposites by conventional polymer processing depends on the strong interfacial interactions between the polymer matrix and the clay. For highly polar polymers such as polyamides (Yano et al., 1993), strong interactions between the clay layers and the polymer increase their contact area and promote the generation of sufficient shear forces. Nonpolar materials, such as PE and PP, however, interact only weakly with mineral surfaces, making the preparation of exfoliated nanocomposites by melt blending challenging. To overcome these problems, chemical modifications of polyolefin (PO) in the form of a coupling agent in PO/clay systems have been used. Functional groups of the coupling agent increased the interactions between the clay surfaces and the polymer chain, providing a favorable enthalpy of mixing clay with the polymer matrix. For PP/clay nanocomposites (Lee, Park, and Sain, 2006; Lee et al., 2011; Zhai et al., 2010; Zhai and Park, 2011; Zhai, Park, and Kontopoulou, 2011; Zheng, Lee, and Park, 2010), PP-g-MAH was usually used as the coupling agent.

In general, the clay and coupling agent contents, as well as the processing conditions, determine the dispersion of clay in the polymer. First, the clay content has a significant effect on the dispersion of clay in PO. At low clay contents, such as 0.02–1 wt%, the exfoliated structure of clay could be obtained in low-density polyethylene (LDPE) (Lee, Wang et al., 2007). At a higher clay loading, the degree of exfoliation decreased where both intercalated and exfoliated structures were present. As the clay content increased to 5 wt%, however, only an intercalated structure could be achieved in LDPE. Second, to exfoliate the clay in PO, high-coupling agent content is often needed. For example, 50 wt% PE-g-MAH could generate a fully exfoliated structure in PE (Lee, Park et al., 2007). Finally, appropriate processing conditions were also important to exfoliate clay in PO. Processing conditions should be set to ensure enough residence time and a moderately high shear force inside the extruder to facilitate the diffusion of polymer chains into the clay layers.

## 3.3 EXTRUSION FOAMING OF PP/CLAY NANOCOMPOSITES

Generally, the experimental setup for extrusion foaming contains the single extrusion system (Park and Cheung, 1997; Zhai et al., 2010) and the tandem extrusion system (Lee, Wang, and Park, 2006). For the tandem extrusion system, as shown in Figure 3.1, the first extruder is used for plasticizing the polymer resin and the second extruder provides mixing and cooling to completely dissolve the $CO_2$ in the

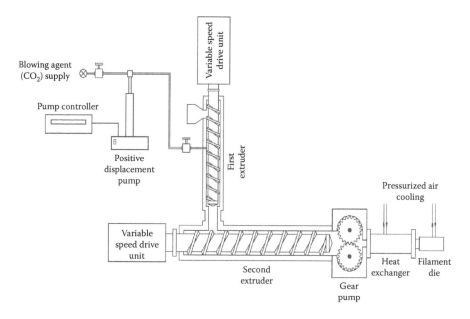

**FIGURE 3.1** Experimental setup of the tandem extrusion foaming system. (From Lee P. C., Wang J., and Park C. B., 2006, *Ind. Eng. Chem. Res.* 45, 175–81.)

polymer melt. Both a chemical blowing agent and a physical blowing agent can be used in the extrusion foaming. When using a chemical blowing agent, the blowing agent needs to be mixed with polymer matrix before foaming. Whereas when using a physical blowing agent, the gas is injected into the barrel during the extrusion process, and the gas content is accurately adjusted and regulated by controlling both the gas flow rate of the syringe pump and the material flow rate that passed through the die. Under the shear and extensional force of the screw, polymer and gas gradually translate from two phase to single phase. When the homogeneous polymer/gas solution enters the filament die, it experiences a rapid pressure drop, which causes a sudden decrease in gas solubility in the polymer; hence, a large number of bubbles are nucleated in the polymer matrix, and a foam structure is finally produced, followed by bubble growth.

The effects of nanoclay addition on PP extrusion foaming have been widely investigated. Table 3.1 summarizes the improvement of the PP foams with the introduction of nanoclay. These results indicated that the presence of nanoclay significantly increases the cell density and decreases the density of the foamed samples. For example, with the addition of 1–5 wt% nanoclay, the cell density increased two to three orders of magnitude and the maximum expansion ratio reached nine times higher than that of neat PP (Zhai et al., 2010). Similar phenomena were obtained in other PP/clay and PP/silica extrusion foaming (Chaudhary and Jayaraman, 2011; Zhai and Park, 2011; Zhai, Park, and Kontopoulou, 2011; Zheng, Lee, and Park, 2010). Moreover, the presence of nanoclay was verified to improve cell morphology by enhancing cell density and to broaden the suitable foaming window, where the suitable foaming window meant that the foams in those foaming regions

**TABLE 3.1**

**The Improvement of PP Foams by Introducing Nanoclay**

| Polymer Nanocomposites | Blowing Agent | Increase of Cell Density[a] (Cells/cm$^3$) | Changing of Expansion Ratio[b] |
|---|---|---|---|
| HPP/clay[c] | $CO_2$ | $2.0 \times 10^7$ versus $1.0 \times 10^9$ | 1.5–2.1 versus 2.3–7.0 |
| HPP/clay[d] | $CO_2$ | $10^{4-5}$ versus $10^{6-8}$ | 1.7–2.2 versus 1.7–18.8 |
| Linear PP/silica[e] | $CO_2$ | $10^{4-5}$ versus $10^{7-9}$ | 2.2–12.9 versus 3.0–19.5 |
| TPO/clay[f] | $CO_2$ | $10^{4-5}$ versus $10^{5-7}$ | 1.5–4.2 versus 1.5–10.4 |
| Linear PP/clay[g] | AC | $10^5$ versus $1.3 \times 10^6$ | 2.2 versus 2.5–3.1 |

[a] The former was the cell density of pure polymer foam, and the latter was that of nanocomposite foam.

[b] The former was the expansion ratio of neat polymer, and the latter was that of polymer nanocomposites.

[c] PP homopolymer: 1–5 wt% clay; 5 wt% $CO_2$ as physical blowing agent.

[d] PP homopolymer: 1–5 wt% clay; 5 wt% $CO_2$ as physical blowing agent.

[e] PP heterophasic copolymer: 1–5 wt% silica; 5 wt% $CO_2$ as physical blowing agent.

[f] PP homopolymer-based soft thermoplastic polyolefin: 0.5–2 wt% clay; 5–50 wt% poly(ethylene-co-octene); 5 wt% $CO_2$ as physical blowing agent.

[g] PP homopolymer: 3–8 wt% clay; 3 wt% AC as chemical blowing agent.

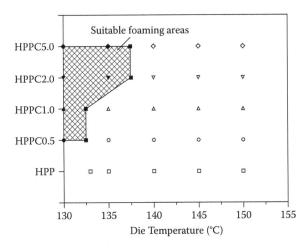

**FIGURE 3.2** Suitable foaming window of HPP and HPPC nanocomposites in fabricating foams with well-defined cell morphology. (From Zhai W. T. et al., 2010, *Ind. Eng. Chem. Res.* 49, 9834–45.)

had well-defined cell structures with very thin walls and uniform cell distributions. As indicated in Figure 3.2, no suitable foaming window was found in pure PP with a high-melt flow rate of 18 g/10 min. With the introduction of nanoclay, however, a suitable foaming window (2.5–7°C) could be obtained in preparing good foamed samples (Zhai et al., 2010).

### 3.3.1 Effect of Clay Dispersion

The effect of clay dispersion on the extrusion foaming behavior of polymer/clay nanocomposites has been widely investigated. Lee, Wang et al. (2007) found that the exfoliated clay greatly improved the cell density in the nanocomposites. When a small amount (< 0.1 wt%) of clay was added into LDPE with PE-g-MAH, the cell density of foams was improved significantly (>$10^9$ cells/cm$^3$). It was found that the optimum content of clay was 0.02–0.1 wt%, which corresponded to an extremely high cell density. With the increase in clay content, however, a gradual decrease in cell density occurred, and no distinct effect of clay on cell density of foams was observed at clay loading of 5 wt%, where LDPE/clay nanocomposite presented an intercalated structure. In the study of PS/clay nanocomposite foaming, Han et al. (2003) quantifiably investigated the influence of clay's dispersion on the cell density of foams. They found that the addition of a small amount (5 wt%) of intercalated nanoclay greatly reduced cell size from 25.3 to 11.1 μm and increased cell density from $2.7 \times 10^7$ to $2.8 \times 10^8$ cells/cm$^3$. When the exfoliated nanoclay was used, the foamed nanocomposite possessed the highest cell density ($1.5 \times 10^9$ cells/cm$^3$) and the smallest cell size (4.9 μm) at the same particle concentration.

For PP/clay (Zhai et al., 2010; Zhai and Park, 2011; Zhai, Park, and Kontopoulou, 2011; Zheng, Lee, and Park, 2010), HDPE/clay (Jo and Naguib, 2007a,b; Lee et al., 2005; Lee et al., 2010) nanocomposite foaming, researchers generally observe that the well dispersed nanoclay increases the cell density by two to three orders of magnitude and greatly improves the cell morphology.

### 3.3.2 Effect of Clay Content

In the study of PP/clay extrusion foaming, researchers usually investigate the effect of clay content on the foaming behavior of PP. In general, the dispersion state of nanoclay, which is related to the nanoclay concentration, can significantly affect the cell density of PP/clay nanocomposite foams. Zheng, Lee, and Park (2010) found that lower clay content (< 0.5 wt%) did not completely suppress cell coalescence of PP/clay foams. With the increase of the clay content, the expansion ratio and cell density of foams were dramatically improved. In cases where the clay content was 0.5 and 1 wt%, the cell coalescence was completely suppressed and the cells were totally closed. When the clay content was higher (5 wt%), however, it could cause a stiffening of the polymer molecular chain that hindered the bubble growth, and finally led to decreases in the expansion ratio of PP foams. Zhai et al. (2010) found that when the die temperature is suitable for foaming, the increased clay contents tended to improve the cell morphology and foaming window of PP/clay foams up to 5 wt%. Similar phenomenon is also observed in LDPE/clay (Lee, Park et al., 2007; Lew, Murphy, and McNally, 2004), PLA/clay (Pilla et al., 2009) nanocomposite foams.

### 3.3.3 Effect of Die Temperature

During the extrusion foaming of PP, an optimal die temperature to maximize the expansion ratio of PP foam is observed for a given nucleating agent content, blowing agent content, and die geometry (Naguib, Park, and Reichelt, 2004; Naguib et al., 2006). In the case of high die temperature, most of the gas escaped through the hot skin layer of foam during expansion because of the high gas diffusivity, the as-prepared PP foams exhibited low expansion ratio. With the decrease of die temperature, the amount of gas lost was decreased because of the reduced diffusivity and the cell coalescence was partially suppressed due to the increased melt strength. Consequently, the increased gas content remained in the cell structure, resulting in the increased volume expansion of PP foams. When the die temperature was too low, the expansion ratio of PP foams was governed by the solidification (i.e., the crystallization) of PP.

In the study of HMSPP foaming blown by butane, Naguib, Park, and Reichelt (2004) discussed the effect of die temperature on volume expansion ratio and cell density of PP foams. The different optimum die temperature that produced the maximum expansion ratio was observed for the different gas content injection. For the cell density of PP foams, however, they did not change with the die temperature, which indicated the decreased die temperature did not induce the enhanced cell nucleation. In the case of PP/clay (Zhai et al., 2010), PP/silica (Zhai, Park, and Kontopoulou, 2011), and nanocomposites extrusion foaming, however, Zhai et al., (2010), found that the decreased die temperature not only increased the expansion ratio of PP foams but also increased their cell densities. This phenomenon resulted from the enhanced cell nucleation and the suppressed cell coalescence, and the mechanisms will be discussed in the next section. In addition, for PP nanocomposite foaming systems, a typical mountain shape of foam expansion relative to the die temperature was also observed.

## 3.4 MECHANISM OF NANOCLAY ON PP FOAMING

### 3.4.1 Nanoclay Enhances Cell Nucleation

A large number of studies have examined the nucleation enhancement mechanism during polymeric foaming processes. As indicated in classical nucleation theory, a lower free energy barrier will result in the formation of bubbles on a heterogeneous nucleating site. In the previous studies, however, the experimental data was not in good quantitative agreement with theoretical predictions without the use of fitting parameters (Taki, 2008). Therefore, the exact mechanism behind cell nucleation is still controversial.

Recently, the importance of shear stress upon cell nucleation has been proposed. The shear stress is induced by expanding gas cavities (Lee, 1993) or by a shear flow field induced during foaming (Guo and Peng, 2003). Han and Han (1990) reached a similar conclusion by observing continuous foaming in situ through transparent slit dies, leading them to suggest that cell nucleation could be induced by flow or shear stress due to the motion of gas clusters, even at thermodynamically unsaturated conditions.

More recently, Leung et al. (2010) and Wang et al. (2010) demonstrated that the melt flow induced by the expansion of previously nucleated bubbles could generate new bubbles around them, as shown in Figure 3.3. The authors speculated that

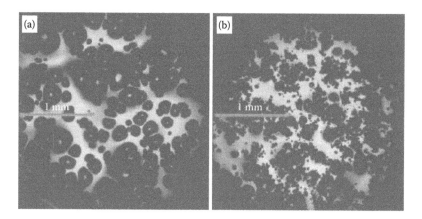

**FIGURE 3.3** Micrographs of PS foaming with 2.1 wt% $CO_2$ at 180°C: (a) pure PS at 2.20 s and (b) PS + 5 wt% talc at 1.56 s. (From Wang C. et al., 2010, *Ind. Eng. Chem. Res.* 49, 12783–92.)

the growing bubbles could generate tensile stress fields around nearby nanofiller particles, resulting in local pressure fluctuations. This discontinuity at the interface between a nucleating agent particle and the surrounding polymer melt yielded local pressure and stress fields around the particle that were different from those in the mixture, which might enhance it as a potential heterogeneous nucleation site. Based on the computer simulation, furthermore, Wang et al. (2010) presented that the pressure profile around nucleating agents could vary significantly from the surroundings, which implied that the assumption of using one system pressure while ignoring any local pressure fluctuation was imprecise.

It is believed that, similar to inorganic filler-enhanced nucleation, the shear/extensional flow in different regions of the die, as well as the expansion of nucleated bubbles near the nanoparticles, would generate a pressure fluctuation around the suspended nanoparticles. The schematic in Figure 3.4 illustrates the induced-extensional flow around the side surface of the nanoclay particle. In extreme situations, such a local pressure field may even be negative and significantly promote cell nucleation. More details about the cell nucleation mechanism of polymer/fillers foaming systems have been presented in our newly submitted journal paper (Zhai et al., 2012), and the readers may get more information from this review paper.

### 3.4.2 NANOCLAY SUPPRESSES CELL COALESCENCE

In general, cell coalescence has a negative effect on the final cell density of foamed samples. During the bubble growth, a biaxial extensional flow is always formed in cell walls. When the cell wall cannot endure any extensional force, it is inclined to rupture and thereby coalesce cells. This often occurs at the later stage of cell growth, as the cells get to contact each other, cell coalescence tends to become severe for polymers that have low melt strength or high processing temperatures. The presence of nanoparticles that are strongly compatible with the polymer matrix can effectively increase the melt strength, especially with the particles well dispersed and

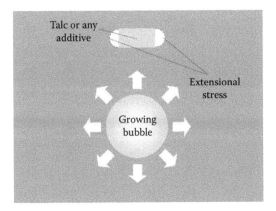

**FIGURE 3.4** Generation of local pressure field around a nanoclay particle by an expanding bubble. (From Leung S. N. et al., 2012, *J Supercrit. Fluids* 63, 187–98.)

aligned (Okamoto, Nam, Maiti, Kotaka, Nakayama et al., 2001), thereby improving the polymer's foaming behaviors.

The orientation of nanoparticles in the polymer matrix is the main reason why nanoparticles can increase the melt strength and induce strain hardening of a polymer. Okamoto, Nam, Maiti, Kotaka, Hasegawa et al. (2001) observed the formation of a house of cards structure in PP/clay nanocomposite melt under elongational flow by TEM analysis. Both strong strain-induced hardening and rheopexy features were originated from the perpendicular alignment of the silicate layers to the stretching direction. Lele et al. (2002) found that compatibilized intercalated PP/clay materials presented high zero-shear viscosity at low shear rates and exhibited a yield behavior at high stresses. Further, rheo-XRD experiments showed that the yield behavior was associated with the orientation of the clay particles. However, the uncompatibilized polymer/clay nanocomposites showed a much smaller zero-shear viscosity and a much less pronounced yield-like behavior. Generally, one expects to observe an orientation if an affine deformation occurs because the uniaxial extensional flow causes strong streamlining along the flow direction. As a result of the streamlining, the silicate layers with a high aspect ratio will preferentially align along the stretching direction (Yuan, Song, and Turng, 2007).

It is well known that cell growth can induce an extensional force along the cell wall. Okamoto, Nam, Maiti, Kotaka, Nakayama et al. (2001) suggested that the induced extensional force during batch foaming was strong enough to align the nanoclays in the cell wall, as indicated in Figure 3.5. Owing to the orientated nanoclays, the improved melt strength of the polymer matrix was able to withstand the stretching force against the thin cell wall. The clay particles seemed to act like a secondary cloth layer, and protected the cells from being destroyed by the external force. In the presence of melt flow, such as during extrusion or injection-molding foaming processes, the shear force, resulting from the screw; and the extensional force, resulting from the die, would be imposed on the nanoparticles (Paul and Robeson, 2008). Consequently, the nanoparticles tended to orient along the flow direction before foaming. It is expected that this orientation would be partially maintained during foaming because the relaxation of

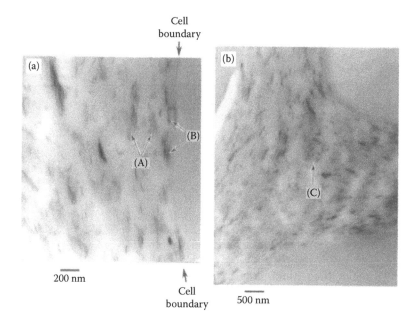

**FIGURE 3.5** TEM micrographs show the orientation of clays in the cell wall during PP/clay nanocomposite batch foaming: (a) one cell wall and (b) junction of three contacting cells. (From Okamoto M., Nam P. H., Maiti P., Kotaka T., Nakayama T. et al., 2001, *Nano. Lett.* 1, 503–5.)

the orientation generally takes more than 1000 s after the cessation of the shear (Lele et al., 2002). The foam processing temperature is typically very low in practical extrusion processes (Park, Behravesh, and Venter, 1998), indicating that a higher degree of nanoparticle orientation might occur for continuous foaming than for batch processing.

## 3.5   INFLUENCE OF FOAMING ON THE DISPERSION OF NANOFILLERS

It is known that the cell growth process results from the extensional flow of the polymer/gas solution, and that the cell walls are biaxially stretched during polymeric foaming (Taki, 2008). A quantitative calculation was carried out to obtain the generated strain rate during cell growth based on an *in situ* visualization foaming analysis (Zhai et al., 2012). The researchers found that the nucleated bubbles are spherical and are of varying sizes, even after a very short nucleation time. As time passes, the bubbles grow and are close to another. A curve fitting was carried out to describe the bubble growth kinetics by treating the smallest detectable bubble. The bubble-growth-applied strain rate on the cell wall, that is, $\dot{\varepsilon}_S$ (in bubble surface direction, characterized by a biaxial extensional process) and $\dot{\varepsilon}_R$ (in bubble diameter direction, characterized by a uniaxial extensional process) was estimated. Furthermore, the estimated $\dot{\varepsilon}_S$ and $\dot{\varepsilon}_R$ were summarized. It is observed that the $\dot{\varepsilon}_S$ rapidly increases from 112.5 s$^{-1}$ to 179.4 s$^{-1}$ at the relative time less than 0.08 s, and then decreases from 178.6 s$^{-1}$ to 82.7 s$^{-1}$ within the followed 0.12 s. In the case of $\dot{\varepsilon}_s$, its value keeps decreasing from 36.3 s$^{-1}$ to 7.4 s$^{-1}$ within 0.20 s.

Given the interface bonding between the polymer matrix and the nanoparticles, the applied biaxial stretching action during cell growth is expected to transfer from the matrix onto the nanoparticles. This process tended to redisperse the nanoparticles in the foamed samples. In the study of linear PP/clay nanocomposites extrusion foaming, Zheng, Lee, and Park (2010) found that the intercalation and exfoliation of the clay particles in PP/clay nanocomposites were improved significantly by the foaming process. Zhai, Park, and Kontopoulou (2011) found that, compared to the unfoamed PP/silica nanocomposites, the size of silica aggregates clearly decreased and were uniformly dispersed in the foamed samples. A similar phenomenon was observed in polycarbonate/silica nanocomposites prepared by batch foaming (Zhai et al., 2006).

## 3.6 CRYSTAL MORPHOLOGY OF PP AND PP/CLAY NANOCOMPOSITE FOAMS

PP is one kind of semicrystalline polymer; its crystallization behavior during the foaming affects the crystal morphology and melting behavior of PP foams. During extrusion foaming, the crystallization behavior of PP foams can be affected in two ways (Zhai et al., 2010). One is the potential for PP to crystallize inside the extruder die before foam expansion. During extrusion processing, the blowing agent takes a plasticizing effect and significantly increases the mobility of PP chains. This can dramatically affect the crystallization kinetics of PP, such as crystallization temperature (Naguib, Park, and Song, 2005), crystallization time, and crystal morphology. Meanwhile, researchers have found that the temperature distribution is nonuniform in the extruder and the polymer flow near the die is extremely low. Once the die temperature is lower than the melting peak of gas plasticized PP, the PP melt near the die at this juncture can crystallize isothermally, and the formed crystals can affect the crystallization behavior of the foamed sample. In addition, high shear and/ or extensional stress field are present in the extruder, which can orient PP chains and significantly affect the crystallization dynamic of the PP resin.

The other affecting factor is PP crystallization outside of the die during foam expansion. It is known that foam expansion is accompanied by a gas cooling process with a very high cooling rate, which will lead to a temperature gradient across the foamed filament. That is, at higher temperature in the core, large-sized crystals will be present in the cell walls because of their longer crystallization time, whereas at lower temperature near the skin, only smaller-sized crystals can be found in the cell walls due to their shorter crystallization time. In addition, it is well known that during the cell growth, cell walls are strongly extended and that the degree of stretching has a significant effect on the crystal size in the cell walls. It should be noted that, however, the degree of stretching, that is, expansion ratio, is related to the die temperature and cooling rate; it is difficult to quantitatively investigate the relationship between the stretching and the crystallization of PP resin.

Due to the enhanced crystal nucleation, the addition of nanoclay dramatically affected the crystal size in the cell walls (Mihai, Huneault, and Favis, 2009). As shown in Figure 3.6, Zhai et al. (2010) found that compared with the pure PP foam crystals

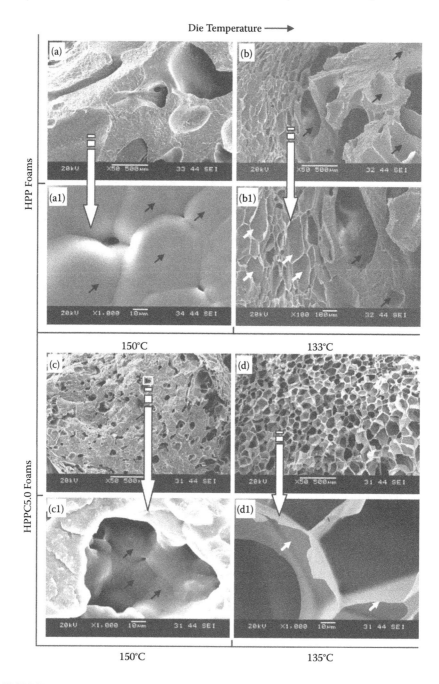

**FIGURE 3.6** The spherulites in cell walls of HPP and HPPC 5.0 foams obtained at various die temperatures. (From Zhai W. T. et al., 2010, *Ind. Eng. Chem. Res.* 49, 9834–45.)

obtained at the same die temperature, the crystals in the PP/clay nanocomposite foam were much smaller. Moreover, it is observed that at a lower die temperature, only small-sized crystals could be observed, even at the foam's core. This is because of the strong stretching force enhancing crystal nucleation and facilitating the increase in crystal density and the decrease in crystal size (Koronfield, Kumaraswamy, and Issaian, 2002).

Furthermore, the extrusion foaming process seems to induce a secondary crystallization. As shown in Figure 3.7 (Zhai et al., 2010), it can be seen that only a single peak is present in the differential scanning calorimetry (DSC) curve of the PP and PP/clay unfoamed samples. It was observed that, however, a weak low melting peak was present in the DSC curve of the PP and PP/clay foamed samples, and the changed heating rate did not affect this peak's existence. This is a local phenomenon involving a section of a given chain and crystallization regions that are limited in the direct vicinity of primary lamellar crystals (Alizadeh et al., 1999). Secondary crystallization occurs after primary crystallization, and the formed crystals exhibit lower perfection compared with those formed during primary crystallization.

It must be pointed out that the crystallization process for crystalline polymer is very complex; the present research is too limited for us to fully understand the crystallization kinetics during the extrusion foaming process. Hence, more work should be done to further quantitatively analyze the crystallization kinetics of PP during the extrusion foaming process.

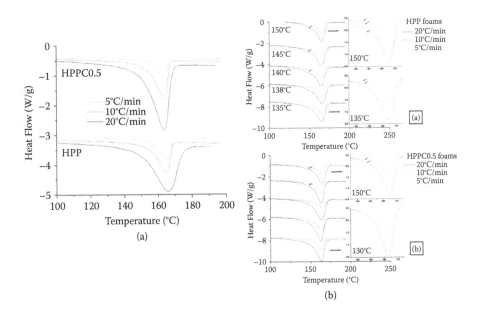

**FIGURE 3.7** DSC curves of (a) unfoamed HPP and HPPC 0.5 samples and (b) foamed HPP and HPPC 0.5 samples measured at heating rates of 5, 10, 20°C/min. (From Zhai W. T. et al., 2010, *Ind. Eng. Chem. Res.* 49, 9834–45.)

## 3.7 CONCLUSIONS

Linear PP has low melt strength and exhibits poor extrusion foaming behavior. In this chapter, the effect of nanoclay addition on the foaming behavior of PP was summarized. In general, the presence of well-dispersed nanoclay enhanced cell nucleation and suppressed cell coalescence, results in a significant increase in cell density by two or three orders of magnitude, and foam expansion. Considering that the dispersion state of nanoclay is connected with the clay concentration, low clay loading around 1 wt% is usually preferable during PP/clay extrusion foaming. The bubble growth is one kind of extensional flow in nature, which tends to redisperse the nanoclay by strain transferring from the PP matrix onto the nanoclay. The crystallization of PP occurs in a complicated atmosphere during foam extrusion; the introduction of nanoclay tends to reduce the perfection of the formed crystals in PP foams.

## REFERENCES

Alizadeh, A., S. Sohn, J. Quinn, and H. Marand. 1999. Influence of structural and topological constraints on the crystallization and melting behavior of polymers. 1. ethylene/1-octene copolymers. *Macromolecules* 32:6221–35.

Burt, J. G. 1978. The elements of expansion of thermoplastics Part II. *J Cell Plast* 14:341–5.

Chaudhary, A. K. and K. Jayaraman. 2011. Extrusion of linear polypropylene–clay nanocomposite foams. *Polym Eng Sci* 51:1749–56.

Danaei, M., N. Sheikh, and F. A. Taromi. 2005. Radiation cross-linked polyethylene foam: Preparation and properties. *J Cell Plast* 41:551–62.

Dennis, H. R., D. L. Hunter, D. Chang, S. Kim, J. L. White, J. W. Cho, and D. R. Paul. 2001. Effect of melt processing conditions on the extent of exfoliation in organoclay-based nanocomposites. *Polymer* 42:9513–22.

Doroudiani, S., C. B. Park, and M. T. Kortschot. 1998. Processing and characterization of microcellular foamed high density polyethylene/isotactic polypropylene blends. *Polym Eng Sci* 38:1205–15.

Guo, M. C. and Y. C. Peng. 2003. Study of shear nucleation theory in continuous microcellular foam extrusion. *Polym Test* 22:705–9.

Guo, G., K. H. Wang, C. B. Park, Y. S. Kim, and G. Li. 2007. Effects of nanoparticles on density reduction and cell morphology of extruded metallocene polyethylene/wood fiber nanocomposites. *J Appl Polym Sci* 104:1058–63.

Han, J. H. and C. D. Han. 1990. Bubble nucleation in polymeric liquids. II. Theoretical considerations. *J Polym Sci Part B: Polym Phy* 28:743–61.

Han, D. H., J. H. Jang, H. Y. Kim, B. N. Kim, and B. Y. Shin. 2006. Manufacturing and foaming of high melt viscosity of polypropylene by using electron beam radiation technology. *Polym Eng Sci* 46: 431–7.

Han, X. M., C. C. Zeng, L. J. Lee, K. W. Koelling, and D. L. Tomasko. 2003. Extrusion of polystyrene nanocomposite foams with supercritical $CO_2$. *Polym Eng Sci* 43:1261–75.

Jo, C. and H. E. Naguib. 2007a. Constitutive modeling of HDPE/clay nanocomposite foams. *Polymer* 48:3349–60.

Jo, C. and H. E. Naguib. 2007b. Effect of nanoclay and foaming conditions on the mechanical properties of HDPE-clay nanocomposite foams. *J Cell Plast* 43:111–21.

Koo, C. M., J. H. Kim, K. H. Wang, and I. J. Chung. 2005. Melt-extensional properties and orientation behaviors of polypropylene-layered silicate nanocomposites. *J Polym Sci B: Polym Phys* 43:158–67.

Koronfield, J. A., G. Kumaraswamy, and A. M. Issaian. 2002. Recent advances in understanding flow effects on polymer crystallization. *Ind Eng Chem Res* 41:6383–92.

Lee, S. T. 1993. Shear effects on thermoplastic foam nucleation. *Polym Eng Sci* 33:418–22.

Lee, S. H., M. Kontopoulou, C. B. Park, A. Wong, and W. T. Zhai. 2011. Optimization of dispersion of nanosilica particles in a PP matrix and their effect on foaming. *Intern Polym Proc* 26:1–11.

Lee, Y. H., T. Kuboki, C. B. Park, M. Sain, and M. Kontopoulou. 2010. The effects of clay dispersion on the mechanical, physical and flame-retarding properties of wood fiber/ polyethylene/clay nanocomposites. *J Appl Polym Sci* 118:452–61.

Lee, Y. H., C. B. Park, and M. Sain. 2006. Strategies for intercalation and exfoliation of PP/ clay nanocomposites. SAE World Congress Paper 06M–109.

Lee, Y. H., C. B. Park, M. Sain, M. Kontopoulou, and W. G. Zheng. 2007. Effects of clay dispersion and content on the rheological mechanical properties, and flame retardance of HDPE/clay nanocomposites. *J Appl Polym Sci* 105:1993–9.

Lee, Y. H., C. B. Park, K. H. Wang, and M. H. Lee. 2005. HDPE-clay nanocomposite foams blown with supercritical $CO_2$. *J Cell Plast* 41:487–502.

Lee, M., C. Tzoganakis, and C. B. Park. 1998. Extrusion of PE/PS blends with supercritical carbon dioxide. *Polym Eng Sci* 38:1112–20.

Lee, P. C., J. Wang, and C. B. Park. 2006. Extruded open-cell foams using two semi-crystalline polymers with different crystallization temperatures. *Ind Eng Chem Res* 45:175–81.

Lee, Y. H., K. H. Wang, C. B. Park, and M. Sain. 2007. Effects of clay dispersion on the foam morphology of LDPE/clay nanocomposites. *J Appl Polym Sci* 103:2129–34.

Lele, A., M. Mackley, G. Galgali, and C. Ramesh. 2002. *In situ* rhro-X-ray investigation of flow-induced orientation in layered silicate syndiotactic polypropylene nanocomposite melt. *J Rheol* 46:1091–110.

Leung, S. N., A. Wong, C. Wang, and C. B. Park. 2012. Mechanism of extensional stress-induced cell formation in polymeric foaming processes with the presence of nucleating agents. *J Supercrit Fluids* 63:187–98.

Lew, C. Y., W. R. Murphy, and G. M. McNally. 2004. Preparation and properties of polyolefin-clay nanocomposites. *Polym Eng Sci* 44:1027–35.

Liu, C., D. Wei, A. Zheng, Y. Li, and H. Xiao. 2006. Improving foamability of polypropylene by grafting modification. *J Appl Polym Sci* 101:4114–23.

Mihai, M., M. A. Huneault, and B. D. Favis. 2009. Crystallinity development in cellular poly(lactic acid) in the presence of supercritical carbon dioxide. *J Appl Polym Sci* 113:2920–32.

Naguib, H. E., C. B. Park, P. C. Lee, and D. L. Xu. 2006. A study on the foaming behaviors of PP resins with talc as nucleating agent. *J Polym Eng* 26:565–87.

Naguib, H. E., C. B. Park, U. Panzer, and N. Reichelt. 2002. Strategies for achieving ultra low-density PP foams. *Polym Eng Sci* 42:1481–92.

Naguib, H. E., C. B. Park, and N. Reichelt. 2004. Fundamental foaming mechanisms governing the volume expansion of extruded polypropylene foams. *J Appl Polym Sci* 91:2661–8.

Naguib, H. E., C. B. Park, and S. W. Song. 2005. Effect of supercritical gas on crystallization of linear and branched polypropylene resins with foaming additives. *Ind Eng Chem Res* 44:6685–91.

Nam, P. H., P. Maiti, M. Okamoto, T. Kotaka, T. Nakayama, M. Takada, M. Ohsima, A. Usuki, N. Hasegawa, and H. Okamoto. 2002. Processing and cellular structure of polypropylene/ clay nanocomposites. *Polym Eng Sci* 42:1907–18.

Okamoto, M., P. H. Nam, P. Maiti, T. Kotaka, N. Hasegawa, and A. Usuki. 2001. A house of cards structure in polypropylene/clay nanocomposites under elongational flow. *Nano Lett* 1:295–8.

Okamoto, M., P. H. Nam, P. Maiti, T. Kotaka, T. Nakayama, M. Takada, M. Ohsima, A. Usuki, N. Hasegawa, and H. Okamoto. 2001. Biaxial flow-induced alignment of silicate layers in polypropylene/clay nanocomposite foam. *Nano Lett* 1:503–5.

Park, C. B., A. H. Behravesh, and R. D. Venter. 1998. Low density microcellular foam processing in extrusion using $CO_2$. *Polym Eng Sci* 38:1812–23.

Park, C. B. and L. K. Cheung. 1997. A study of cell nucleation in the extrusion of polypropylene foams. *Polym Eng Sci* 37:1–10.

Park, J. U., J. L. Kim, D. H. Kim, K. H. Ahn, and S. J. Lee. 2006. Rheological behavior of polymer/layered silicate nanocomposites under uniaxial extensional flow. *Macromol Res* 14:318–23.

Paul, D. R. and L. M. Robeson. 2008. Polymer nanotechnology: Nanocomposites. *Polymer* 49:3187–204.

Pilla, S., S. G. Kim, G. K. Auer, S. Q. Gong, and C. B. Park. 2009. Microcellular extrusion-foaming of polylactide with chain-extender. *Polym Eng Sci* 49:1653–60.

Rachtanapun, P., S. E. M. Selke, and L. M. Matuana. 2003. Microcellular foam of polymer blends of HDPE/PP and their composites with wood fiber. *J Appl Polym Sci* 88:2842–50.

Rachtanapun, P., S. E. M. Selke, and L. M. Matuana. 2004. Effect of the high-density polyethylene melt index on the microcellular foaming of high-density polyethylene/polypropylene blends. *J Appl Polym Sci* 93:364–71.

Ray, S. S. and M. Okamoto. 2003. Polymer/layered silicate nanocomposites: A review from preparation to processing. *Prog Poly Sci* 28:1539–1641.

Reichelt, N., M. Stadlbauer, R. Folland, C. B. Park, and J. Wang. 2003. PP-blends with tailored foamability and mechanical properties. *Cell Polym* 22:315–27.

Rodríguez-Pérez, M. A. 2005. Crosslinked polyolefin foams: Production, structure, properties, and applications. *Adv Polym Sci* 184:97–126.

Ruinaard, H. 2006. Elongational viscosity as a tool to predict the foamability of polyolefins. *J Cell Plast* 42:207–20.

Shen, J., C. C. Zeng, and L. J. Lee. 2005. Synthesis of polystyrene-carbon nanofibers nanocomposite foams. *Polymer* 46:5218–24.

Spitael, P. and C. Macosko. 2004. Strain hardening in polypropylenes and its role in extrusion foaming. *Polym Eng Sci* 44:2090–100.

Taki, K. 2008. Experimental and numerical studies on the effects of pressure release rate on number density of bubbles and bubble growth in a polymeric foaming process. *Chem Eng Sci* 63:3643–53.

Vasile, C. and R. B. Seymour. 1993. *Handbook of Polyolefins*. New York: Marcel Dekker.

Wang, C., S. N. Leung, M. Bussmann, W. T. Zhai, and C. B. Park. 2010. Numerical investigation of nucleating agent-enhanced heterogeneous nucleation. *Ind Eng Chem Res* 49:12783–92.

Wong, A., R. K. M. Chu, S. N. Leung, and C. B. Park. 2011. A batch foaming visualization system with extensional stress-inducing ability. *Chem Eng Sci* 66:55–63.

Yano, K., A. Usuki, A. Okada, T. Kurauchi, and O. Kamigaito. 1993. Synthesis and properties of polyimide-clay hybrid. *J Polym Sci Part A: Polym Chem* 31:2493–8.

Yuan, M. J., Q. Song, and L. S. Turng. 2007. Spatial orientation of nanoclay and crystallite in microcellular injection molded polyamide-6 nanocomposites. *Polym Eng Sci* 47:765–79.

Yuan, M. J., L. S. Turng, S. Q. Gong, D. Caulfield, C. Hunt, and R. Spindler. 2004. Study of injection molded microcellular polyamide-6 nanocomposites. *Polym Eng Sci* 44:673–86.

Zeng, C. C., X. M. Han, L. J. Lee, K. W. Koelling, and D. L. Tomasko. 2003. Polymer–clay nanocomposite foams prepared using carbon dioxide. *Adv Mater* 15:1743–7.

Zhai, W. T. 2012. Nanoparticle addition improves the foamability of polypropylene. *SPE Plastics Research Online* (DOI: 10.1002/spepro.004263).

Zhai, W. T., T. Kuboki, L. Wang, C. B. Park, E. K. Lee, and H. E. Naguib. 2010. Cell structure evolution and the crystallization behavior of polypropylene/clay nanocomposites foams blown in continuous extrusion. *Ind Eng Chem Res* 49:9834–45.

Zhai, W. T. and C. B. Park. 2011. Effect of nanoclay addition on the foaming behavior of linear PP-based soft TPO foam blown in continuous extrusion. *Polym Eng Sci* 51:2387–97.

Zhai, W. T., C. B. Park, and M. Kontopoulou. 2011. Nanosilica addition dramatically improves the cell morphology and expansion ratio of polypropylene heterophasic copolymer foams blown in continuous extrusion. *Ind Eng Chem Res* 50:7282–9.

Zhai, W. T., S. N. Leung, L. Wang, T. Kuboki, and C. B. Park. 2012. The effect of nanoparticles on the foaming behavior of polyolefin-based nanocomposites: A critical review of fundamental issues. *Compos Sci Tech,* submitted.

Zhai, W. T., J. Wang, N. Chen, H. E. Naguib, and C. B. Park. 2012. The orientation of carbon nanotubes in poly(ethylene-co-octene) microcellular foaming and its suppression effect on cell coalescence. *Polym Eng Sci* 52:2078–89.

Zhai, W. T., H. Y. Wang, J. Yu, J. Y. Dong, and J. S. He. 2008a. Cell coalescence suppressed by cross-linking structure in polypropylene microcellular foaming. *Polym Eng Sci* 48:1312–21.

Zhai, W. T., H. Y. Wang, J. Yu, J. Y. Dong, and J. S. He. 2008b. Foaming behavior of isotactic polypropylene in supercritical $CO_2$ influenced by phase morphology via chain grafting. *Polymer* 49:3146–56.

Zhai, W. T., H. Y. Wang, J. Yu, J. Y. Dong, and J. S. He. 2008c. Foaming behavior of polypropylene/polystyrene blends enhanced by improved interfacial compatibility. *J Polym Sci B: Polym Phys* 46:1641–51.

Zhai, W. T., J. Yu, L. C. Wu, W. M. Ma, and J. S. He. 2006. Heterogeneous nucleation uniformizing cell size distribution in microcellular nanocomposites foams. *Polymer* 47:7580–9.

Zheng, W. G., Y. H. Lee, and C. B. Park. 2006. The effects of exfoliated nano-clay on the extrusion microcellular foaming of amorphous and crystalline nylon. *J Cell Plast* 42:271–88.

Zheng, W. G., Y. H. Lee, and C. B. Park. 2010. Use of nanoparticles for improving the foaming behaviors of linear PP. *J Appl Polym Sci* 117:2972–9.

# 4 Foams Based on Starch, Bagasse Fibers, and Montmorillonite

*Suzana Mali, Fabio Yamashita, and Maria Victoria E. Grossmann*

## CONTENTS

## 4.1   INTRODUCTION

Polymer foams, among polymers in film, sheet, or molded forms, have been widely used for packaging, which generates a large amount of household waste that is difficult to collect and to recycle (Cinelli et al., 2006; Zhou, Song, and Parker, 2006). On the other hand, there has been increasing interest in the research and development of starch-based materials to provide biodegradable alternatives for packaging, because starch is obtained from renewable resources, is low cost, and is abundant.

Starch foams with insulating properties similar to those of polystyrene foam have been industrially produced by extrusion or baking processes. There are several patents describing the production of starch foams by extrusion (Bastioli et al., 1994, 1998a, 1998b; Bellotti et al., 1995, 2000; Lacourse and Altieri, 1989, 1991; Xu and Doane, 1997, 1998).

Foaming extrusion has mainly been used to produce loose-fill packaging materials, with several advantages, such as the ability to process high-viscosity polymers in the absence of solvents, large operational flexibility due to the broad range of processing conditions (0–500 atm and 70–500°C), the feasibility of multiple injections, and control of both residence time (distribution) and the degree of mixing (Liu et al., 2009; van Duin, Machado, and Covas, 2001).

Baked starch foams have been largely studied by several authors (Chiellini et al., 2009; Cinelli et al., 2006; Guan and Hanna, 2006; Kaisangsri, Kerdchoechuen, and Laohakunjit, 2012; Salgado et al., 2008; Shogren, Lawton, and Tiefenbacher, 2002;

Soykeabkaew, Supaphol, and Rujiravanit, 2004; Vercelheze et al., 2012). The properties of baked starch foam products will vary with moisture content, starch type, and additives used in the dough formulations (Liu et al., 2009). The foam baking is a simple process that includes two steps: starch gelatinization and water evaporation, which expands the mixture and forms foam, and foam dehydration until a final moisture content of 2–4% is obtained (Shogren et al., 1998). Figures 4.1a,b show the scanning electron micrographs of cassava starch baked foam in contact with the top and with the bottom surfaces of the mold, respectively. It can be observed that the top surface of the tray shows a smooth surface (Figure 4.1a), while the bottom surface presents large pores (Figure 4.1b), which are related to the water evaporation during the baking process. During the baking process, the first contact of the mixture is with the bottom surface of the mold, and probably the water evaporation begins from bottom to top, resulting in such structures observed by scanning electron microscopy.

An alternative method based on a microwave foaming process was described by Zhou, Song, and Parker (2006), which involves converting starch-based raw materials into pellets by extrusion processing, and foaming the extruded pellets by microwave heating. The microwaveable starch pellets are compact for transportation

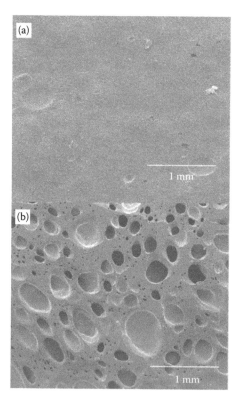

**FIGURE 4.1** Scanning electron micrographs of cassava starch baked foam. (a) Tray surface in contact with the top surface of the mold and (b) tray surface in contact with the bottom surface of the mold. (From authors; data not published.)

and storage, and can be expanded using microwave when needed. They may be formulated to produce microwaveable snacks in the food industry. In nonfood applications, free-flowing foamed balls may be produced for loosefill packaging.

Generally, extruded or baked starch foams present unsatisfactory physical and mechanical properties, such as poor mechanical properties and high water solubility, making these products sensitive to the relative humidity at which they are stored and used (Liu et al., 2009; Mali et al., 2010). The development of new low-cost foams with better performance can be achieved by the incorporation of less hydrophilic polymers and/or other materials that act as reinforcements (fibers, nanoparticles, etc.). Thus, in this chapter we will discuss some alternatives, such as fiber and nanoclay incorporation into starch foams to improve the performance of these materials.

### 4.1.1 STARCH-FIBER COMPOSITE FOAMS

Composite is a multiphase material formed from a combination of two or more materials, which differ in composition or form, but remain bonded together and retain their identity and properties. These materials maintain an interface between components and act to improve specific or synergistic characteristics not obtainable by any of the original components acting alone (Yu, Dean, and Li, 2006).

Foam composites produced from a starch matrix reinforced with natural fibers have been studied in several applications by many authors (Bénézet et al., 2012; Cinelli et al., 2006; Guan and Hanna, 2006; Kaisangsri, Kerdchoechuen, and Laohakunjit, 2012; Lee et al., 2009; Mali et al., 2010; Robin et al., 2011; Salgado et al., 2008; Shogren et al., 1998; Soykeabkaew, Supaphol, and Rujiravanit, 2004; Vercelheze et al., 2012).

Lignocellulosic fibers are abundant in nature and several types of them can be used as a reinforcement in polymeric matrices, such as jute, sisal, bamboo, cotton, ramie, flax, curauá, and banana. Some others are agro-industrial or agricultural residues, such as sugarcane and cassava bagasses; rice; oat and soy hulls; barley and wheat straws, which also have a great potential to be used for this purpose.

Natural fibers obtained from agro-industrial or agricultural residues (or bagasse fibers) have some attractive properties when compared to inorganic fibers, such as lower cost and lower density, which is interesting for foam production. In addition, the use of these materials can reduce the need for burning or decomposing these wastes, and also opens up new markets for the agricultural industry (Ardanuy, Antunes, and Velasco, 2012; Mishra and Sain, 2009).

Natural fibers also have advantages from the point of view of fiber–matrix adhesion, specifically with polar matrix materials, such as starch matrices. Cellulose is the main component of fibers, for example, sugarcane bagasse consists of 32–44% cellulose, banana fiber has 60–65%, sisal and ramie consist of about 65 and 83% of cellulose, respectively (Satyanarayana, Arizaga, and Wypych, 2009). As cellulose is a polyhydroxylated macromolecule, these hydroxyl groups can form hydrogen bonds inside the macromolecule itself, between other cellulose macromolecules (Habibi et al., 2008), and with other polyhydroxylated macromolecules, such as starch. According to Guan and Hanna (2006), when more cellulose was introduced into starch materials, more hydrogen bonds were formed among starch and cellulose chains making the blend more crystalline. Lee et al. (2009) reported that the presence of fibers resulted in good bonding with the starch, which can form a stronger matrix, thereby increasing the compressive strength of the foam.

Shogren et al. (1998) produced starch foams using a baking process. The authors reported that the addition of softwood fibers increased starch foams' strength when they were stored in low and high relative humidities. Lawton, Shogren, and Tiefenbacher (2004) found that the addition of 5–10% fiber clearly produced higher strength foams because the fibers adhered well to the starch matrix, and thus acted as reinforcement.

Soykeabkaew, Supaphol, and Rujiravanit (2004) produced starch-based composite foams by baking starch batters incorporating either jute or flax fibers. Mechanical properties (flexural strength and flexural modulus of elasticity) were markedly improved with the addition of 5–10% of the fibers. The authors attributed the results to the strong interaction between the fibers and the starch matrix.

Cinelli et al. (2006) reported that potato starch-based trays could be foamed with a relatively high content of corn fibers (28.9 to 54.7%), however, the fibers' addition did not exert a reinforcing effect when used at the higher content.

Salgado et al. (2008) produced baked foams based on cassava starch, sunflower proteins, and cellulose fibers (eucalypt pulps), and they reported that increments of fiber concentration from 10 to 20% improved the mechanical properties but increased the water absorption capacity of the material in at least 15%.

Debiagi et al. (2010) investigated the production of foam composites by extrusion, from the mixture cassava starch, glycerol (plasticizer), and two different types of natural fibers (oat hulls and sugarcane bagasse). The composites were prepared in a single screw extruder, with three different levels of each fiber (0, 5, and 10 g/100 g starch), two levels of humidity (18 and 26%), and a fixed level of glycerol (20 g/100 g solids). The addition of fibers reduced the water solubility index of the foams.

Mali et al. (2010) obtained extruded foams made from cassava starch, sugarcane bagasse fibers, and polyvinyl alcohol (PVA) and stressed that 20% (w/w) of the fiber addition improved the water resistance, but 40% of the fiber addition resulted in denser and more colored foams, with a yellowish aspect.

Schmidt and Laurindo (2010) employed cassava starch, dolomitic limestone, and eucalypt cellulose fibers to prepare foam trays to pack foodstuffs. They investigated the influence of the cellulose fiber concentration (5, 10, 15, 20, 30, and 40%) in the composite formulation. The results indicated that an increase in cellulose fiber concentration promoted a decrease in density and tensile strength of the foam samples. The tensile strength at break for foam trays containing 5% of cellulose fibers was 3.03 MPa, while the commercial trays of expanded polystyrene used to pack foods in supermarkets presented a tensile strength of 1.49 MPa. The elongation at break of the foam trays varied slightly with the increase in cellulose fiber concentration, the values being about 20% lower than the elongation at break observed for commercial foam trays of expanded polystyrene.

Debiagi et al. (2011) produced foams made from cassava starch, polyvinyl alcohol (PVA), sugarcane bagasse fibers, and chitosan by extrusion. The composites were prepared with formulations determined by a constrained ternary mixtures experimental design, using as variables: ($X_1$) starch/PVA (100–70%), ($X_2$) chitosan

(0–2%), and ($X_3$) fibers from sugarcane (0–28%). As reported by the authors, fiber addition at intermediary levels (14%) improved the expansion and mechanical properties of the foams, and when fibers were added at higher levels (28%), expansion index and mechanical properties decreased. There was a trend of red and yellow colors when the composites were produced with the highest proportion of fibers. Figure 4.2 shows some SEM micrographs of these foams; foams produced exclusively with starch/PVA (without fiber addition) resulted in a material with good expansion, which was observed by the opened cell structure (Figure 4.2a) when compared to the foams produced with high-fiber proportions (28%), which showed the closed cell structure (Figure 4.2b). According to Moraru and Kokini (2003) and Preechawong et al. (2004), the opened cell structure of extruded foams was a result of the venting of a large amount of water molecules when the starchy polymer emerged from the extruder die.

In Figure 4.2b, we can also observe a fiber accumulation (arrow) on the extruded foam, resulting in a nonhomogeneous structure (Figure 4.2). Carr et al. (2006) related this nonhomogeneity of fiber distribution to a decrease in the compression strength of the foams, which was also observed by Debiagi et al. (2011).

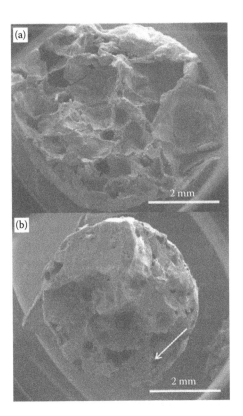

**FIGURE 4.2** Scanning electron micrographs of extruded foams formulated as follows: (a) 100% starch/PVA and (b) 70% starch/PVA, 2% chitosan, and 28% fibers. (From Debiagi F. et al., 2011, *Brazilian Archives of Biology and Technology* 54, 1043–1052.)

Baked foam trays based on cassava starch blended with 4% chitosan and 30% of kraft fibers presented some good properties, such as density, tensile strength, and elongation of 0.14 g/cm³, 944.40 kPa, and 2.43%, respectively, but water absorption and water solubility indexes that were greater than those of polystyrene foam (Kaisangsri, Kerdchoechuen, and Laohakunjit, 2012).

Benézét et al. (2012) reported that the addition of fibers increased the expansion index and led to a reduction in water adsorption of extruded potato starch foams. They also reported that the fiber additions at a 10% level generally improved the starch foam's properties.

In the last few years there has been an increasing interest in studying another class of fillers that can be used alone or in combination with the fibers to produce starch foam composites, and although several nanoparticles have been recognized as possible additives to enhance polymer performance, the most intensive studies are currently focused on layered silicates, such as montmorillonite (MMT), due to their availability, versatility, low cost, and respectability toward the environment and health (Azeredo, 2009; Vercelheze et al., 2012).

Some patents have reported that the introduction of fibers and/or inorganic fillers is interesting for improving mechanical properties of starch materials (Andersen et al., 1998; Andersen and Hodson, 1995, 2001). Other studies have shown that it is possible to obtain food packaging from mixtures of starch, fibers, water, and other additives by thermopressing or baking (Carr et al., 2006; Schmidt and Laurindo, 2010), and these products could be an alternative to the use of expanded polystyrene foams (Vercelheze et al., 2012).

## 4.2  STARCH-MONTMORILLONITE NANOCOMPOSITE FOAMS

Nanocomposites are systems that contain fillers with at least one nanosized dimension and represent a new class of materials that exhibit improved mechanical, thermal, barrier, and physicochemical properties compared with the starting polymers and conventional (microscale) composites (Azeredo, 2009).

The montmorillonite crystal lattice consists of 1-nm thin layers with an octahedral alumina sheet sandwiched between two tetrahedral silica sheets. The layers are negatively charged, and this charge is balanced by alkali cations, such as $Na^+$, $Li^+$, or $Ca^{2+}$, in the gallery space between the aluminosilicate layers. Na-montmorillonite (Na-MMT) clay is hydrophilic with a high surface area and is miscible with hydrophilic polymers, such as starch (Ardakani et al., 2010; Ray and Okamoto, 2003; Vercelheze et al., 2012).

In contrast with the tactoid structure predominating in microcomposites (conventional composites), in which the polymer and the clay tactoids remain immiscible, resulting in agglomeration of the clay in the matrix and poor macroscopic properties of the material (Alexandre et al., 2009; Ludueña, Alvarez, and Vasquez, 2007), the interaction between layered silicates and polymer chains may produce two types of ideal nanoscale composites. The properties of the resulting material are dependent on the state of the nanoclay in the nanocomposite, that is, if it is exfoliate or intercalate. Intercalation is the state in which polymer chains are present between the clay layers, resulting in a multilayered structure with alternating polymer/inorganic layers. Exfoliation is the state in which the silicate layers are completely separated and dispersed in a continuous polymer matrix (Weiss, Takhistov, and McClements, 2006). The exfoliated nanocomposites involve extensive polymer penetration, with the clay layers delaminated and randomly dispersed in the polymer matrix (Ludueña, Alvarez, and Vasquez, 2007). Exfoliated nanocomposites have been reported to exhibit the best properties due to the optimal interaction between clay and polymer (Adame and Beall, 2009; Alexandre et al., 2009; Osman, Rupp, and Suter, 2005).

Nanoclays have also been reported to improve the mechanical strength of biopolymers and this is related to the strong interfacial interaction between the polymeric matrix and clay, which change the morphology of the polymeric matrix (Avella et al., 2005; Cyras et al., 2008; Park et al., 2003). Wilhelm et al. (2003), Cyras et al. (2008), and Moraes, Muller, and Laurindo (2012) have all reported the increase of mechanical strength in starch–nanoclay films. Thus, a very low loading of MMT (1–10%) results in a large improvement in mechanical, thermal, electrical, and barrier properties of the nanocomposites (Liu et al., 2011; Xiong et al., 2008).

As reported by several authors, the dispersion of nanofiller within the polymer matrix is a key step to obtain exfoliated structures (Lee and Hanna, 2008; Liu et al., 2011). Turri, Alborghetti, and Levi (2008) reported that to produce a true nanocomposite, the clay stack must first be delaminated within the polymer matrix. Delamination of the clay then gives rise to a homogeneous dispersion of individual platelets.

According to Le Corre, Bras, and Dufresne (2010),

> … there are several techniques for preparing such materials. For all techniques there are two steps: mixing and processing, which often occur at the same time. Processing methods are usually the same as for pure polymers: extrusion, injection molding, and casting or compression molding. However, special attention must be brought to the processing temperature when working with organic fillers. The choice of the matrix depends on several parameters such as the application, the compatibility between components, the process, and the costs.

To improve their dispersibility, clays could be modified with organic surfactants, which are typically the quaternary ammonium salts of long fatty acid chains. These surfactants decrease the surface tension of the aluminosilicate particulates, which in turn reduces the endothermal enthalpy of mixing. Many organophilic nanoclays or organoclays, therefore, have already been studied, and some of their products are already marketed on an industrial scale (Matsuda et al., 2012; Park et al., 2002; Turri, Alborghetti, and Levi, 2008).

Lee, Chen, and Hanna (2008) described the preparation of extruded foams based on cassava starch, poly(lactic acid) (PLA), and four different organoclays (Cloisite®10A, Cloisite 25A, Cloisite 93A, and Cloisite 15A). The authors reported that the first X-ray diffraction peaks for all four nanocomposite foams were observed to shift to lower angles compared to those of the original organoclays, indicating that the intercalation of starch/PLA polymer into the organoclay layers occurred. They also reported that the organoclays had compatible interactions with the starch/PLA molecules, resulting in decreased water absorption solubility of the samples.

Debiagi and Mali (2012) produced intercalated nanocomposite foams based on cassava starch, polyvinyl alcohol (PVA), and sodium montmorillonite (Na-MMT).

The nanocomposites were prepared in a single-screw extruder using different starch contents (97.6–55.2 g/100 g formulation), PVA (0–40 g/100 g formulation), and Na-MMT (0–4.8 g/100 g formulation), and glycerol was used as a plasticizer. The addition of Na-MMT resulted in an increase of expansion index and mechanical strength of the foams. Na-MMT addition also resulted in a decrease of water absorption capacity of the samples.

Matsuda et al. (2012) developed biodegradable trays based on cassava starch and two different organoclays (Cloisite 10A and 30B) using a baking process and investigated the effects of these components on the microstructural, physicochemical, and mechanical properties of the trays. The ammonium cations of the organoclays were methyl tallow bis (2-hydroxyethyl) and dimethyl benzyl hydrogenated-tallow ammonium for the Cloisite 30B and the Cloisite 10A, respectively. All formulations resulted in well-shaped trays, similar to polystyrene ones, as shown in Figure 4.3, with densities between 0.2809 and 0.3075 g/cm$^3$.

Matsuda et al. (2012) also described the microstructure of the produced foams (Figure 4.4), the scanning electron micrographs of the tray cross sections show the sandwich-type structure of the foams; this structure includes dense outer skins that enclose small cells (i.e., the surface of the foam). The interior of the foams had large cells with thin walls; this type of structure has also been reported by Cinelli et al. (2006) in their study on foams of potato starch and corn fiber. As observed in Figure 4.4, the samples produced with the 5% nanoclay addition (Figure 4.4b,c) showed larger air cells than the starch-based foam (Figure 4.4a); probably the addition of the nanoclays improved the foaming ability of starch pastes, resulting in cell walls that were more resistant to collapse during water evaporation than that which occurred during the baking process.

The advantages of fibers or MMT addition on starch polymeric matrices are discussed in the literature, but the combination of these two fillers to produce starch foams is still poorly investigated. The combination of the biodegradability and low cost of starch and fibers with the strength and stability of nanoclays could be an alternative for new biodegradable materials, which have improved mechanical properties and storage stability.

## 4.3 STARCH-FIBER-MONTMORILLONITE NANOCOMPOSITE FOAMS

Vercelheze et al. (2012) investigated the use of a baking process to prepare composite and nanocomposite trays based on cassava starch, sugarcane bagasse fiber, and Na-MMT. All formulations resulted in well-shaped trays with densities between 0.1941 and 0.2966 g/cm$^3$. The addition of fibers and Na-MMT resulted in less dense and less rigid trays. The foams had high water absorption capacities (>50%) when immersed in water (1 min). The studied processing conditions resulted in good nanoclay dispersion, leading to the formation of an exfoliated structure.

To produce exfoliated nanocomposites, the mixing step is very important. Matsuda et al. (2012) and Vercelheze et al. (2012) employed the same processing conditions to obtain baked nanocomposite foams and reported that the production process favored the exfoliation of MMT. They reported that the raw materials were

**FIGURE 4.3**   Foam trays based on: (a) cassava starch; (b) 95% cassava starch and 5% Cloisite 10A, and (c) 95% cassava starch and 5% Cloisite 30B. (From authors; data not published.)

mixed for 20 min at 18000 rpm. In the first 10 min, water, starch, nanoclays, fibers, and other solids were mixed; glycerol was then added, and the paste was mixed for 10 more min. According to Chung et al. (2010), good dispersion of clays in a starch matrix can be achieved by first preparing the nanocomposites in diluted aqueous solutions followed by plasticization.

When exfoliated nanocomposites are analyzed by X-ray diffraction, the loss of the characteristic peak from the clay can be observed. In Figure 4.5, typical X-ray diffraction patterns of exfoliated nanocomposites were observed; cassava starch foams

**FIGURE 4.4** Scanning electron micrographs of foam trays cross sections based on: (a) cassava starch; (b) 95% cassava starch and 5% Cloisite 10A, and (c) 95% cassava starch and 5% Cloisite 30B. (From Matsuda D. N. K. et al., 2012, *Industrial Crops and Products*, http://dx.doi.org/10.1016/j.indcrop.2012.08.032.)

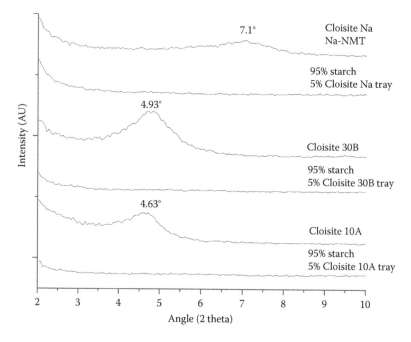

**FIGURE 4.5** X-ray diffraction patterns of foam trays based on cassava starch and MMT. (Adapted from Vercelheze A. E. S. et al., 2012, *Carbohydrate Polymers* 87, 1302–1310 and Matsuda D. N. K. et al., 2012, *Industrial Crops and Products*, http://dx.doi.org/10.1016/j. indcrop.2012.08.032.)

were produced with three different types of MMT (Cloisite Na, Cloisite 10A, and Cloisite 30B), and in all foams (Figure 4.5) there was no diffraction peak between $2\theta = 4°$ and $2\theta = 10°$. The loss of the peak from the clay is due to the good clay dispersion, which leads to disordered clay tactoids with a low concentration of clay agglomerates and indicates the formation of an exfoliated structure.

The major challenge involved in the research and development of new biodegradable foams based on starch is to produce products that are competitive in cost and performance with expanded polystyrene, and the use of fillers such as natural fibers and nanoclays can allow many technological and environmental issues to be solved; the properties of these new products will vary with process technology, starch type, and filler addition in the foam formulations.

## REFERENCES

Adame, D. and Beall, G. W. 2009. Direct measurement of the constrained polymer region in polyamide/clay nanocomposites and the implications for gas diffusion. *Applied Clay Science* 42:545–552.

Alexandre, B., Langevin, D., Médéric, P., Aubry, T., Couderc, H., Nguyen, Q. T., Saiter, A., and Marais, S. 2009. Water barrier properties of polyamide 12/montmorillonite nanocomposite membranes: Structure and volume fraction effects. *Journal of Membrane Science* 328:186–204.

Andersen, P. J., Christiansen, B. J, Hodson, S. K., and Shaode, O. 1998. *Composites for Manufacturing Sheets Having a High Starch Content*. U.S. Patent 5.976.235.

Andersen, P. J., and Hodson, S. K. 1995. *Methods for Uniformly Dispersing Fibers within Starch-Based Compositions*. U.S. Patent 5.618.341.

Andersen, P. J., and Hodson, S. K. 2001. *Thermoplastic Starch Compositions Incorporating a Particulate Filler Component*. U.S. Patent 6.231.970.

Ardakani, M. M., Mohseni, M. A., Beitollahi, H., Benvidi, A., and Naeimi, H., 2010. Electrochemical determination of vitamin C in the presence of uric acid by a novel $TiO_2$ nanoparticles modified carbon paste electrode. *Chinese Chemical Letters* 21:1471–1474.

Ardanuy, M., Antunes, M., and Velasco, J. I. 2012. Vegetable fibres from agricultural residues as thermo-mechanical reinforcement in recycled polypropylene-based green foams. *Waste Management* 32:256–263.

Avella, M., De Vlieger, J. J., Errico, M. E., Vacca, P., and Volpe, M. G. 2005. Biodegradable starch/clay nanocomposite films for food packaging applications. *Food Chemistry* 93:467–474.

Azeredo, H. M. C. 2009. Nanocomposites for food packaging applications. *Food Research International* 42:1240–1253.

Bastioli, C., Bellotti, V., Del Giudice, L., Lombi, R., and Rallis, A. 1994. *Expanded Articles of Biodegradable Plastic Materials*. U.S. Patent 5,360,830.

Bastioli, C., Bellotti, V., Del Tredici, G., Montino, A., and Ponti, R. 1998a. *Biodegradable Foamed Plastic Materials*. U.S. Patent 5,736,586.

Bastioli, C., Bellotti, V., Del Tredici, G., and Rallis, A. 1998b. *Biodegradable Foamed Articles and Process for Preparation Thereof*. U.S. Patent 5,801,207.

Bellotti, V., Bastioli, C., Rallis, A., and Del Tredici, G. 1995. *Expanded Articles of Biodegradable Plastic Material and a Process for the Preparation Thereof*. Europe Patent, EP0667369.

Bellotti, V., Bastioli, C., Rallis, A., and Del Tredici, G. 2000. *Expanded Articles of Biodegradable Plastic Material and a Process for the Preparation Thereof*. Europe Patent, EP0989158.

Bénézet, J. C., Stanojlovic-Davidovic, A., Bergeret, A., Ferry, L., and Crespy, A. 2012. Mechanical and physical properties of expanded starch, reinforced by natural fibres. *Industrial Crops and Products* 37:435–440.

Carr, L. G., Ponce, P., Parra, D., Lugao, A. B., and Buchler, P. M. 2006. Influence of fibers on the mechanical properties of starch-foams based for thermal pressed products. *Journal of Polymers and the Environment* 14:179–183.

Chiellini, E., Cinelli, P., Ilieva, V. I., Imam, S. H., and Lawton, J. L. 2009. Environmentally compatible foamed articles based on potato starch, corn fiber, and poly (vinyl alcohol). *Journal of Cellular Plastic* 45:17–32.

Chung Y., Ansari, S., Estevez, L., Hayrapetyan, S., Giannelis, E. P., and Lai, H. M. 2010. Preparation and properties of biodegradable starch–clay nanocomposites. *Carbohydrate Polymers* 79:391–396.

Cinelli, P., Chiellini, E., Lawton, J. W., and Imam, S. H. 2006. Foamed articles based on potato starch, corn fibers and poly (vinyl alcohol). *Polymers Degradation and Stability* 91:1147–1155.

Cyras, V. P., Manfredi, L. B., Ton-That, M. T., and Vázquez, A. 2008. Physical and mechanical properties of thermoplastic starch/montmorillonite nanocomposite films. *Carbohydrate Polymers* 73:55–63.

Debiagi, F. and Mali, S. 2012. Functional properties of extruded nanocomposites based on cassava starch, polyvinyl alcohol and montmorillonite. *Macromolecular Symposia* 319:235–239.

Debiagi, F., Mali, S., Grossmann, M. V. E., and Yamashita, F. 2010. Efeito de fibras vegetais nas propriedades de compósitos biodegradáveis de amido de mandioca produzidos via extrusão. *Ciência e Agrotecnologiav* 34:1522–1529.

Debiagi, F., Mali, S., Grossmann, M. V. E., and Yamashita, F. 2011. Biodegradable foams based on starch, polyvinyl alcohol, chitosan and sugarcane fibers obtained by extrusion. *Brazilian Archives of Biology and Technology* 54:1043–1052.

Guan, J. and Hanna, M. A. 2006. Selected morphological and functional properties of extruded acetylated starch–cellulose foams. *Bioresource Technology* 97:1716–1726.

Habibi, Y., El-Zawawy, W. K., Ibrahim, M. M., and Dufresne, A. 2008. Processing and characterization of reinforced polyethylene composites made with lignocellulosic fibers from Egyptian agro-industrial residues. *Composites Science and Technology* 68:1877–1885.

Kaisangsri, N., Kerdchoechuen, O., and Laohakunjit, N. 2012. Biodegradable foam tray from cassava starch blended with natural fiber and chitosan. *Industrial Crops and Products* 37:542–546.

Lacourse, N. L. and Altieri, P. A. 1989. *Biodegradable Packaging Material and the Method of Preparation Thereof.* U.S. Patent 4,863,655.

Lacourse, N. L. and Altieri, P. A. 1991. *Biodegradable Shaped Products and the Method of Preparation Thereof.* U.S. Patent 5,043,196.

Lawton, J. W., Shogren, R. L., and Tiefenbacher, K. F. 2004. Aspen fiber addition improves the mechanical properties of baked cornstarch foams. *Industrial Crops and Products* 19:41–48.

Le Corre, D., Bras, J., and Dufresne, A. 2010. Starch Nanoparticles: A Review. *Biomacromolecules* 11:1139–1153.

Lee, S. Y., Chen, H., and Hanna, M. 2008. Preparation and characterization of tapioca starch–poly (lactic acid) nanocomposite foams by melt intercalation based on clay type. *Industrial Crops and Products* 28:95–106.

Lee, S. Y., Eskridge, K. M., Koh, W. Y., and Hanna, M. A. 2009. Evaluation of ingredient effects on extruded starch-based foams using a supersaturated split-plot design. *Industrial Crops and Products* 29:427–436.

Lee, S. Y. and Hanna, M. A. 2008. Tapioca starch-poly (lactic acid)-Cloisite 30B nanocomposite foams. *Polymer Composites* 30:665–672.

Liu, H., Chaudhary, D., Yusa, S., and Tadé, M.O. 2011. Glycerol/starch/Na+ montmorillonite nanocomposites: A XRD, FTIR, DSC and 1H NMR study. *Carbohydrate Polymers* 83, 1591–1597.

Liu, H., Xie, F., Yu, L., Chen, L., and Li, L. 2009. Thermal processing of starch-based polymers. *Progress in Polymer Science* 34:1348–1368.

Ludueña, L. N., Alvarez, V. A., and Vasquez, A. 2007. Processing and microstructure of PCL/clay nanocomposites. *Materials Science and Engineering: A* 460:121–129.

Mali, S., Debiagi, F., Grossmann, M. V. E., and Yamashita, F. 2010. Starch, sugarcane bagasse fibre and polyvinyl alcohol effects on extruded foam properties: A mixture design approach. *Industrial Crops and Products* 32:353–359.

Matsuda, D. N. K., Vercelheze, A. E. S., Carvalho, G. M., Yamashita, F., and Mali, S. 2012. Baked foams of cassava starch and organically modified nanoclays. *Industrial Crops and Products*, http://dx.doi.org/10.1016/j.indcrop.2012.08.032.

Mishra, S. and Sain, M. 2009. Commercialization of wheat straw as reinforcing filler for commodity thermoplastics. *Journal of Natural Fibers* 6:83–97.

Moraes, J. O., Muller, C. M. O., and Laurindo, J. B. 2012. Influence of the simultaneous addition of bentonite and cellulose fibers on the mechanical and barrier properties of starch composite-films. *Food Science and Technology International* 18:35–45.

Moraru, C. I. and Kokini, J. L. 2003. Nucleation and expansion during extrusion and microwave heating of cereal foods. *Comprehensive Reviews in Food Science and Food Safety* 2:147–165.

Osman, M. A., Rupp, J. E. P., and Suter, U. W. 2005. Effect of non-ionic surfactants on the exfoliation and properties of polyethylene-layered silicate nanocomposites. *Polymer* 46:8202–8209.

Park, H. W., Lee, W. K., Park, C. Y., Cho, W. J., and Ha, C. S. 2003. Environmentally friendly polymer hybrids. Part I—Mechanical, thermal and barrier properties of thermoplastics starch/clay nanocomposites. *Journal of Material Science* 38:909–915.

Park, H. M., Li, X., Jin, C. Z., Park, C. Y., Cho, W. J., and Ha, C. K. 2002. Preparation and properties of biodegradable thermoplastic starch/clay hybrids. *Macromolecular Materials and Engineering* 287:553–558.

Preechawong, D., Pessan, M., Rujiravanit, R., and Supaphol, P. 2004. Preparation and properties of starch/poly (vinyl alcohol) composite foams. *Macromolecular Symposia* 216:217–227.

Ray, S. S. and Okamoto, M. 2003. Polymer/layered silicate nanocomposites: A review from preparation to processing. *Progress in Polymer Science* 28:1539–1641.

Robin, F., Dubois, C., Pineau, N., Schuchmann, H. P., and Palzer, S. 2011. Expansion mechanism of extruded foams supplemented with wheat bran. *Journal of Food Engineering* 107:80–89.

Salgado, P. R., Schmidt, V. C., Ortiz, S. E., Mauri, A. N., and Laurindo, J. B. 2008. Biodegradable foams based on cassava starch, sunflower proteins and cellulose fibers obtained by a baking process. *Journal of Food Engineering* 85:435–443.

Satyanarayana, K. G., Arizaga, G. G. C., and Wypych, F. 2009. Biodegradable composites based on lignocellulosic fibers—An overview. *Progress in Polymer Science* 34:982–1021.

Schmidt, V. C. R. and Laurindo, J. B. 2010. Characterization of foams obtained from cassava starch, cellulose fibres and dolomitic limestone by a thermopressing process. *Brazilian Archives of Biology and Technology* 53:185–192.

Shogren, R. L., Lawton, J. W., Doanne, W. M., and Tiefenbacher, F. K. 1998. Structure and morphology of baked starch foams. *Polymer* 39:6649–6655.

Shogren, R. L., Lawton, J. W., and Tiefenbacher, K. F. 2002. Baked starch foams: Starch modifications and additives improve process parameters, structure and properties. *Industrial Crops and Products* 16:69–79.

Soykeabkaew, N., Supaphol, P., and Rujiravanit, R. 2004. Preparation and characterization of jute and flax reinforced starch-based composite foams. *Carbohydrate Polymers* 58:53–63.

Turri, S., Alborghetti, L., and Levi, M. 2008. Formulation and properties of a model two-component nanocomposite coating from organophilic nanoclays. *Journal of Polymers Research* 15:365–372.

van Duin, M., Machado, A. V., and Covas, J. 2001. A look inside the extruder: Evolution of chemistry, morphology and rheology along the extruder axis during reactive processing and blending. *Macromolecular Symposium* 170:29–39.

Vercelheze, A. E. S., Fakhouri, F. M., Dall'antônia, L. H., Urbano, A., Youssef, A. E., Yamashita, F., and Mali, S. 2012. Properties of baked foams based on cassava starch, sugarcane bagasse fibers and montmorillonite. *Carbohydrate Polymers* 87:1302–1310.

Weiss, J., Takhistov, P., and McClements, D. J. 2006. Functional materials in food nanotechnology. *Journal of Food Science* 71:R107–R116.

Wilhelm, H. M., Sierakowskia, M. R., Souza, G. P., and Wypychc, F. 2003. Starch films reinforced with mineral clay. *Carbohydrate Polymers* 52:101–110.

Xiong, H. G., Tang, S. W., Tang, H. L., and Zou, P. 2008. The structure and properties of a starch-based biodegradable film. *Carbohydrate Polymers* 71:263–268.

Xu, W. and Doane, W. M. 1997. *Biodegradable Polyester and Natural Polymer Compositions and Expanded Articles There from.* U.S. Patent 5,665,786.

Xu, W. and Doane, W. M. 1998. *Biodegradable Polyester and Natural Polymer Compositions and Expanded Articles There from.* U.S. Patent 5,854,345.

Yu, L., Dean, K., and Li, L. 2006. Polymer blends and composites from renewable resources. *Progress in Polymer Science* 31:576–602.

Zhou, J., Song, J., and Parker, R. 2006. Structure and properties of starch-based foams prepared by microwave heating from extruded pellets. *Carbohydrate Polymers* 63: 466–75.

# 5 Processing of Polymer Nanocomposite Foams in Supercritical $CO_2$

*Sebastien Livi and Jannick Duchet-Rumeau*

## CONTENTS

## 5.1 INTRODUCTION

Since the 1980s, academic and industrial research has had a growing interest in the processing of polymer/filler nanocomposites due to their excellent barrier, electrical, and mechanical properties, which are conferred by the unique advantages of nanoparticles such as their high surface area, aspect ratio, shape, and size. More recently, the market for lightweight materials which are very present in the industry is oriented toward the development of polymer nanocomposite foams in order to improve the compressive properties, mechanical strength, surface quality, thermal behavior, and dimensional stability of unmodified foams. Moreover, current environmental constraints on the protection of ozone has pushed governments to replace the blowing agents commonly used in the industry, such as chlorofluorocarbon (CFC), with other environmentally friendly agents. For these reasons, supercritical $CO_2$

(ScCO$_2$) has emerged an as excellent alternative to chlorofluorocarbon and represents a promising element of green chemistry. Indeed, ScCO$_2$ is a low cost process, which is readily recyclable and nonflammable, with an ability to plasticize many polymers such as fluoropolymers and polysiloxanes, which are known for being soluble in supercritical carbon dioxide. In addition, CO$_2$ has a low surface tension (close to zero), viscosity, and density like a liquid which gives it a high solvency power tunable by adjusting pressure but also a high diffusivity like a gas.

To achieve lightweight polymer nanocomposite foams with high performance without sacrificing mechanical properties, many methods and approaches have been reported in the literature. In this chapter, the recent progress in this field is reviewed. As the nanocomposite foams cover wide scientific fields which extend from materials science—nanoparticles and polymer—to processing of nano-composites and foams and to the characterization of morphology and properties, this chapter is divided into four parts. In the first part, we describe the foaming mechanism under supercritical CO$_2$ of nanocomposite materials by identifying the nature of the interactions between CO$_2$ and matrix as well as the influence of the variation of physical parameters (temperature, pressure) on foaming. In the second part, a brief description of existing foaming processes is introduced. In the third part, the modification of the rheological properties of polymer by the use of organic or inorganic additives such as chain extenders, ionic liquid, and fillers will be pre-sented before describing the influence of nanoparticles on the morphology of foams. Finally, the last part will focus on the structure–properties relationships of polymer nanocomposite foams.

## 5.2 USE OF CARBON DIOXIDE IN SUPERCRITICAL CONDITIONS FOR NANOCOMPOSITE FOAM PREPARATION

### 5.2.1 FORMATION MECHANISM OF POLYMER/NANOFILLER FOAMS

The foaming of polymer is a complex process involving some thermodynamic events occurring in a rheologically evolving media. Just like crystallization, foam-ing consists of a cell nucleation followed by a growth step, and ends with a thermal structural stabilization. Nucleation is initiated by a thermodynamical instability, which occurs in a polymer saturated with supercritical gas at high pressure. Such instability can be driven by a pressure drop or by an increase in the temperature leading to a metastable supersaturated solution. In order to restore a lower energy state, a phase separation occurs between the CO$_2$ and the polymer. In the case of the ternary blend polymer/nanofillers/supercritical CO$_2$, the theory of heteroge-neous nucleation provides a deeper insight about the parameters governing this first step of the foam processing. The heterogeneous nucleation takes place because gas bubbles are formed at the interface between the liquid and the solid phase (Colton and Suh, 1987).

Nucleation rate is thus defined by the following equation:

$$N_{het} = f_1 C_1 \exp\left(\frac{-\Delta G_{het}^*}{kT}\right) \tag{5.1}$$

Where $N_{het}$ is the cell nucleation rate, $C_1$ is the concentration of nucleation sites, $f_1$ is the frequency factor for gas molecules joining the nucleus, $k$ is the Boltzmann's constant, and $T$ is the system temperature. $\Delta G$ is the free energy barrier to initiate heterogeneous cell nucleation, which is given by:

$$\Delta G^*_{het} = \frac{16\pi\gamma_{bp}^3}{3\Delta P^2} S(\theta) \text{ which is also equal to } \Delta G^*_{het} = \Delta G^*_{hom} S(\theta)$$

where $\gamma_{bp}$ is the surface energy of the polymer–bubble interface, $\Delta P$ is the pressure drop of the gas/polymer solution, and $S(\theta)$ depends on the contact angle at the polymer–nucleating agent interface. For a typical contact angle of 20°, $S(\theta)$ is about $10^{-3}$. The energetic barrier for heterogeneous nucleation is significantly reduced with respect to one for homogeneous nucleation. The minimization of the free energy barrier for nucleation increases the cell nucleation rate. Many relevant factors can affect the cell nucleation rate and are included in the material properties and the thermodynamic parameters. As an example, a higher-pressure drop of the saturated polymer would lower the free energy barrier to initiate nucleation, resulting in a higher nucleation rate. Park et al. reported that the pressure drop rate can also increase the nucleation rate (Nalawade, Picchioni, and Janssen, 2006). The size, the shape, and the distribution of particles like the surface treatment can impact the efficiency of nucleation (Han, Zheng, and Lee, 2003). Once nucleation is initiated, the gas diffuses from the polymer matrix to the cell. Bubble expansion is the main mechanism to lower the supersaturation state, which causes a gas concentration decrease in the bubble and promotes its diffusion from the polymer matrix. The solubility and diffusivity of gas in polymer will depend on the polymer or filler/gas chemical affinity and on the pressure and temperature. As the bubble continues to expand, the concentration gradient across the polymer and gas boundary decreases. The structural expansion is therefore hindered by the polymer rheology—which is reinforced during expansion by the cooling and the deplasticization induced by the gas depletion of the polymer matrix. The structure expansion is then stopped when the sample reaches a temperature below the glass transition, Tg. It is important to note that Tg can be hugely decreased due to the presence of the remaining blowing agent in the polymer matrix. The gas uptake will imply modifications of the material's thermal and rheological properties.

## 5.2.2 EFFECT OF FOAMING PARAMETERS ON FOAM MORPHOLOGY

In the majority of the literature, the reduction of the foaming temperature is often presented as a means to increase the cell nucleation and to reduce the cell size (Arora, Lesser, and McCarthy, 1998; Beckmann and Goel, 1994). The increase in the nanoclay amount introduced within nanocomposite can also counterbalance the effect of temperature on the viscosity reduction. The dissolution of the blowing agent such as $CO_2$ in the polymer causes its plasticization. This phenomenon induces a decrease of the glass transition temperature and of the melting point of the polymer (Wissinger and Paulaitis, 1991) and thereby results in a reduction of its viscosity (Kwag, Manke, and Gulari, 1999). Two distinct mechanisms are involved in this material softening. First, the presence of a low molecular mass component dilutes the entanglement of the molten polymer just like a solvent. The second mechanism is the creation of additional free volume, which improves the chain mobility just like an increase in temperature (Tai et al., 2010). This last parameter is the driving force for the viscosity reduction. The required pressure for doping polymer with gas is also acting on the viscosity independently of its effect on the $CO_2$ sorption ratio. Thus, it is possible to consider the concomitant effects of temperature, pressure, and $CO_2$ concentration on the polymer viscosity. In a recent study, Tai et al. (2010) showed that the viscosity of a low molar mass poly (DL-lactic acid) doped with $CO_2$ at 35°C under 10 MPa was similar to the nonplasticized melt one at 140°C (Wolff et al., 2011). The depressurization rate is another foaming parameter that plays a significant effect on the foam morphology. When the depressurization rate is slower, the synthesized foam is more heterogeneous. The cell size is all the more reduced and the cell number is all the more high as the depressurization rate is fast (Han, Zheng, and Lee, 2002).

## 5.3  PROCESSING OF NANOCOMPOSITE FOAMS

Currently, the two most common pathways for producing nanocomposite foams are the batch or continuous process depending on the application targeted and on the scale of the production. The first method is the most commonly used at laboratory scale whereas the second process based on extrusion and injection are more likely compatible with industrial productions since it allows direct end-product shaping (e.g., sheet extrusion, profile extrusion, injection molding).

### 5.3.1  BATCH FOAMING

Nanocomposite foaming in a batch reactor is a simple process that needs the processing of polymer/fillers blended in a previous step before being introduced in an autoclave working at elevated temperature and pressure (Corre et al., 2011a; Jin et al., 2001; Ngo et al., 2010; Stafford, Russell, and McCarthy, 1999). The reactor may be heated by electric heaters or immersed in an oil bath. The pressure is set from a gas cylinder with the help of a pump system. The depressurization is the most critical step of the foaming process since it is responsible for the thermodynamic instability that may induce the phenomena of nucleation and growth of cells. The autoclaves are thus equipped with a pressure release valve that can be manually or automatically activated for a precise and reproducible pressure drop. A batch foaming system is schematically represented in Figure 5.1.

The batch foaming takes place in two steps. The first one is the impregnation of the polymer/nanofillers blend with the blowing agent that can be performed either in the solid state or in the liquid state since the dissolved $CO_2$ tends to bring down the glass transition temperature or the melting temperature of the polymer (DeSimone et al., 1994; Mawson et al., 1995; McHugh et al., 2002). Depending on the foaming route, the autoclave can be heated or cooled down before the initiation of nucleation and the process conditions—constant control of temperature and pressure—play a key role in the foaming process.

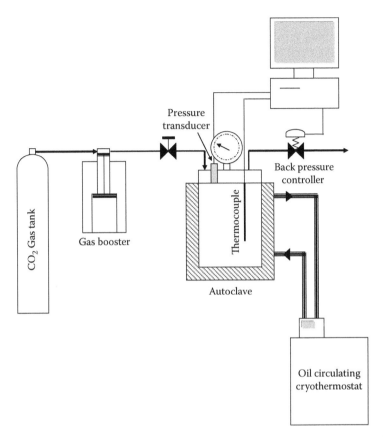

**FIGURE 5.1** Batch foaming process. (Reprinted from Corre Y. M. et al., 2011a, *Journal of Supercritical Fluids* 58, 177–188. Copyright 2013, with permission from Elsevier.)

Many studies have shown that the cell size decreases while the cell number density increases when the pressure is high and the temperature reduced (Corre et al., 2011a). These results are explained by a supersaturation of dissolved $CO_2$ in these conditions of temperature and pressure. For this reason, a quenching process pressure has been developed in recent years to induce foaming and to obtain a good control of the porosity (Jiang et al., 2009; Mascia et al., 2006; Reignier, Gendron, and Champagne, 2007). Another method mentioned in the literature but more difficult to control is based on an increase of the system temperature (Alavi, Rizvi, and Harriott, 2003b; Matuana, 2008). Indeed, after that the sample is impregnated with $CO_2$, the depressurization is performed very slowly at a temperature below the glass transition temperature of the polymer matrix before taking the material out of the reactor and dipping it in a hot oil bath during a given residence time.

However, the batch process has several limitations. The main disadvantage is that this method requires extremely long cycle times for reaching the gas sorption equilibrium due to its low diffusivity through a significant amount of material. Another drawback is linked to the vessel geometry (Corre et al., 2011a). Depending on the volume and the thermal inertia of the reactor, a rapid depressurization at a low temperature may result in freezing of the sample. This sudden cooling of the polymer could block cell expansion.

In conclusion, batch foaming is mainly suitable for the foaming of small amounts of material. A larger volume of material induces additional costs of the supercritical facilities, partly because of the complexity of sealing parts and equipment for gas compression. Most of industrial foam conversions are then conducted either by injection or extrusion foaming.

## 5.3.2 CONTINUOUS MODE

For industrial applications, significant R&D resources have been devoted to develop technology in continuous mode to produce microcellular and lightweight foams of polymer with improved acoustic and insulation properties without sacrificing the mechanical properties for diverse applications, those of food packaging, biomedical devices, or automotive parts (Alavi, Rizvi, and Harriott, 2003a; Arora, Lesser, and McCarthy, 1998; Corre et al., 2011a; DeSimone et al., 1994; Han, Zheng, and Lee, 2002; Jacobsen and Pierick, 2000; Jiang et al., 2009; Jin et al., 2001; Kwag, Manke, and Gulari, 1999; Mascia et al., 2006; Matuana, 2008; Mawson et al., 1995; McHugh et al., 2002; Michaeli and Heinz, 2000; Ngo et al., 2010; Reignier, Gendron, and Champagne, 2007; Stafford, Russell, and McCarthy, 1999; Tai et al., 2010; Wang et al., 2012; Wissinger and Paulaitis, 1991; Wolff et al., 2011). In recent years, various techniques have emerged based on injection or extrusion foaming adapted to polymers with low and/or high viscosity (Jacobsen and Pierick, 2000, 2001; Michaeli and Heinz, 2000).

### 5.3.2.1 Injection Foaming

Currently, one of the best ways to prepare foams suitable for the industry is injection molding directly at the die outlet of the extruder. As a result, injection foaming is a semicontinuous process since it includes a plasticizer/gas extruder and a transfer chamber as depicted in Figure 5.2. The processing of nanocomposite can be performed in the extruder by adding the fillers within the plasticized polymer/gas mixture.

This process must be used for the molding of complex and lightweight parts (i.e., foam density of 0.2 $g/cm^3$). Indeed, this process allows (i) working at relatively low temperatures, (ii) leads to a reduction of the viscosity of the polymeric material due to the plasticizer effect of supercritical $CO_2$, (iii) causes an important decrease of the cycle time, and (iv) preserves the dimensional stability of the final parts. Co-injection can also be used to fill volumetric parts (e.g., car interior panel, heat insulated electronic housing) with foamed material while the surface is made of a massive polymer skin to keep a certain mechanical resistance. To produce microcellular foams, only three technologies of injection molding are used comprising Optifoam by Sulzer Chemtech AG (Switzerland) (Sulzer), ErgoCell by Demag (Germany) (Sauthof, 2003), and MuCell by Trexel Inc. (United States) (Błędzki et al., 2006). The high cost of equipment limits its use in the research laboratories and does not facilitate the development of academic works on theoretical and experimental models that would be useful for a better understanding of the foaming process.

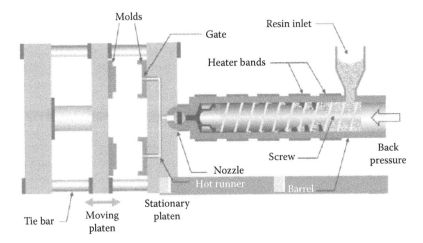

**FIGURE 5.2 (See color insert.)** Injection foaming process. (Reprinted from Lim L. T. et al., 2008, *Progress in Polymer Science* 33, 820–852. Copyright 2013, with permission from Elsevier.)

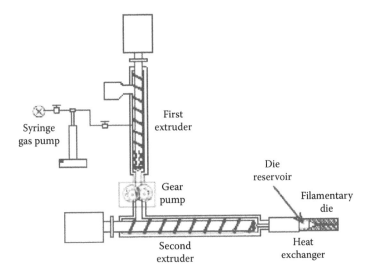

**FIGURE 5.3** Extrusion foaming process. (Reprinted from Alavi S. H. et al., 2003b, *Food Research International* 36, 309–19. Copyright 2013, with permission from Elsevier.)

### 5.3.2.2 Extrusion Foaming

Extrusion foaming by direct gasing is a widespread process for various applications ranging from foamed sheets to expanded profile extrusion with foam density ranging from 0.05 up to 0.5 g/cm$^3$. Extrusion foaming is a melt processing where polymer is first thermomechanically plasticized before being dynamically mixed with a blowing agent. The extrusion process is schematically presented in

Figure 5.3. The addition of nanofillers can be performed either in the melt step or in the mixing step with the gas. The following steps consist of a thermal stabilization of the mixture before shaping the melt through a specific die and foaming by a rapid pressure quench. This specific extrusion can be performed either in mono screw or twin screw extruder and tandem extruders are even reported as an efficient cooling potential if required for the mixture stabilization step. The postextrusion process consists of the solidification of the cellular structure while an ageing process is driven by osmotic gas exchanges between bubbles and air. Depending on the polymer permeation for blowing agent and air, this ageing phenomenon can last from a few hours up to several months depending on the storage conditions and on the blowing agent used. Extrusion foaming is a process where diluted gas acts as a metastable plasticizer. For the difference with batch foaming, the molten polymer/gas solution is here subjected to shear and elongational strains. On one hand, such mechanical solicitations favor the mixing efficiency, but on the other hand, too aggressive conditions (e.g., high screw rotation rates, sudden change in barrel section, flow obstacles...) can induce gas demixing at an early stage for polypropylene foaming with $CO_2$.

Recently, research has intensified efforts in the production of poly(lactic acid) (PLA) foams from extrusion (Lee, Kareko, and Jung, 2008; Sauceau et al., 2011). In fact, different authors investigated the foaming of amorphous and commercially available polylactid polymer matrices with high carbon dioxide amount (9–10%wt) in order to understand the $CO_2$ foaming. Generally, foams with high open-cell content and poor mechanical properties have been obtained (Reignier, Gendron, and Champagne, 2007; Sauceau et al., 2011). The use of semicrystalline PLA or the addition of fillers (starch, nanoclay) induces an increase of viscosity reducing the cell size and strengthening the PLA mechanical properties (Mihai et al., 2007; Wang et al., 2012).

However, although this technique has many advantages, it is necessary to equip an extruder with a system of injection of supercritical $CO_2$ and with a static mixer, which can lead to a significant cost. Then, several precautions must be taken to avoid phase separation before nucleation such as (i) the estimation of the residence time of the filled polymer/$CO_2$ mixture in the extruder (using theoretical models developed in the literature); (ii) the amount of $CO_2$ injected that is less than the solubility thereof, and (iii) the control of the depressurization of the mixture through the nozzle to lead to a saturation of the $CO_2$, which generates the production of microcellular foams. Park et al. have demonstrated that the pressure-induced solubility change can be employed as a continuous microcellular nucleation mechanism by using a rapid pressure drop element consisting of a nozzle (Park and Suh, 1993). In 1995, the same author had also studied the effect of pressure drop rate on cell nucleation and the cell density by using nozzles with different radius (Park, Baldwin, and Suh, 1995). Thus, they concluded that the pressure drop rate in the nucleation device have a key role to determine the cell density of the extruded foam. In fact, an increase of the drop pressure rate leads to an improvement of the cell density. More recently, many studies have focused on the use of supercritical $CO_2$ as foaming agents in the field of food industry (Alavi et al., 1999; Alavi, Rizvi, Harriott, 2003a). Indeed, for starch-based microcellular foams, different mathematic models have been investigated toward a better understanding of the mechanisms of bubble growth and collapse but also to demonstrate the importance of material parameters on the optimization of the foaming process (Alavi, Rizvi, Harriott, 2003a).

In summary, very few theoretical and experimental studies have been devoted to the development of mathematical models to simulate, predict, and characterize the pressure distribution in different matrices, or the size distribution of bubbles. In conclusion, it is absolutely necessary to improve the current simulations to provide a better understanding of the foaming process and hopefully a growing interest in the industry for large scale production.

## 5.4 STRUCTURE–PROPERTY RELATIONSHIP OF POLYMER NANOCOMPOSITE FOAMS

If processing parameters (temperature-pressure-soaking time-depressurization rate) have a clear impact on the foam morphology, materials parameters can also be tuned to tailor the foams structuration. The rheological properties of polymer material studied and in particular, the elongational viscoelasticity which is responsible for the structure and cell size obtained, must be well characterized. For this reason, the introduction into the polymer matrix of chemical or physical heterogeneities is required to increase the viscosity of polymer and to limit cell growth (Doroudiani, Park, and Kortschot, 1998; Pilla et al., 2009; Ray and Okamoto, 2003a). In order to create microcellular polymeric foams, two pathways have been commonly studied: (i) the use of different additives to increase the melt strength of the matrix and to limit cell growth, or (ii) the addition of organic/inorganic fillers such as silicates, silica, and metal oxides. In addition, a high interface amount and the multiplication of nucleation sites generated by the use of nanoparticles is a huge advantage.

### 5.4.1 Effect of Chain Extender on Foam Morphology

With the objective of improving the viscoelastic properties, many works involving chain extension and chain branching were reported in the literature on biodegradable polymers (Corre et al., 2011b; Gedler et al., 2012; Mihai, Huneault, and Favis, 2010; Park, Behravesh, and Venter, 1997; Pilla et al., 2009; Stange and Munstedt, 2006; Yang et al., 2012; Zeng et al., 2010; Zhou et al., 2007). For example, in the case of PLA matrix, the chain extension consists of reacting the carboxyl or hydroxyl groups of PLA chains with functional groups such as amine, anhydride, isocyanate, hydroxyl, or epoxy of chain extender (Reverchon and Cardea, 2007). The use of 1 to 2% of chain extender was demonstrated to be sufficient for branching the PLA matrix but also to allow continuous foaming by extrusion (Mihai, Huneault, and Favis, 2010; Zhou et al., 2007). Recently, Corre et al. (2011b) have demonstrated that the chain extension improved the viscosity, the shear sensitivity of the molten PLA, as well as the elastic properties of the matrix. They have determined one density of $10^{10}$ cells/cm$^3$ on high chain extended PLA instead of $10^4$ cells/cm$^3$ on neat PLA for the same foaming temperature (110°C). Indeed, as the viscoelasticity in the melt state is considerably increased by the chain extension process, the potential for cell expansion is greatly reduced. Other works have corroborated these results in the literature (Reverchon and Cardea, 2007; Stange and Munstedt, 2006; Tsivintzelis, Angelopoulou, and Panayiotou, 2007). Moreover, these branched polymers display a strain hardening behavior, which is a requirement for melt extensional processes such as blowing or foaming (Gendron and Daigneault, 2000; Meissner and Hostettler, 1994; Nam, Yoo, and Lee, 2005; Spitael, 2004). The use of these chain extenders increases the crystallinity of the polymer, which can have an effect on the morphology of the foams (Corre et al., 2011a). According to the literature, the presence of melt heterogeneities due to the coexistence of a crystalline phase and an amorphous phase results in an inhomogeneous distribution of adsorbed $CO_2$ in the polymer matrix (Corre et al., 2011a,b; Gendron and Daigneault, 2000; Meissner and Hostettler, 1994; Nam, Yoo, and Lee, 2005; Reverchon and Cardea, 2007; Spitael, 2004; Stange and Munstedt, 2006; Tsivintzelis, Angelopoulou, and Panayiotou, 2007). In fact, the adsorption of $CO_2$ does not occur in the crystalline phase, which indicates a reduction of the $CO_2$ (Baldwin, Park, and Suh, 1996a,b; Doroudiani, Park, and Kortschot, 1996). These heterogeneities of fluid density may cause the creation of a liquid-like phase if the foaming is carried out above the critical conditions (Tatibouet and Gendron, 2005). Several authors have concluded that the increased rate of crystallization has a significant impact on the foaming process as well as a strengthening of the viscoelastic properties of the polymer (Mascia et al., 2006; Zhai, Yu, Wu et al., 2006). Several authors have shown that for polymers PCL, PBS, PLLA, or PC, an increase in the $CO_2$ pressure leads to a decrease of the crystallization temperature of polymers (Lian et al., 2006; Naguib, Park, and Song, 2005; Takada, Hasegawa, and Ohshima, 2004; Zhai, Yu, Ma et al., 2006). This phenomenon is accentuated in branched polymers (Naguib, Park, and Song, 2005; Zhai, Yu, Ma et al., 2006).

In conclusion, the use of a chain extender can be very useful to limit the cell's growth while improving the viscoelastic properties of polymers. A compromise must be found between crystallinity, desired rheological properties, and foaming parameters.

### 5.4.2 Effect of Ionic Liquids as New Additives on Foam Morphology

Ionic liquids (ILs) are other innovative additives that can be introduced within a polymer matrix (Livi, Gérard, and Duchet-Rumeau, 2011a,b; Livi et al., 2013; Soares et al., 2012). In recent works, the addition of small amounts of ILs (1 wt%) to a polystyrene (PS) matrix has played a significant role in the preparation of microcellular foams at low temperature-pressure conditions in supercritical medium (Livi et al., 2013). The cell morphologies have been tuned by the wide choice of cation–anion combinations including pyridinium, imidazolium, and phosphonium cations functionalized either with long alkyl chains considered as $CO_2$-phobe or perfluorinated ones that are known to be $CO_2$-philic. Thus, the use of conventional ionic liquids generates a microcellular morphology more coarse than one of pure polystyrene (120–130 μm compared to 80 μm for neat PS). SEM micrographs on fractured specimens of microcellular PS/IL foams are shown in Figure 5.4.

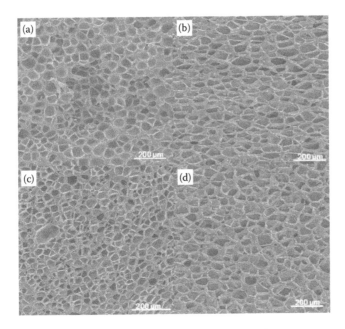

**FIGURE 5.4** (a) SEM micrographs of neat PS. (b) Blends PS/C18Py I. (c) PS/C18C18Im I. (d) PS/C18P I.

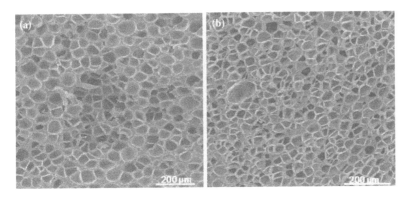

**FIGURE 5.5** (a) SEM micrographs of neat PS. (b) Blends C18C12FIm I.

In contrast, the use of fluorinated ionic liquids generates lower polarizability and solubility parameters. Thus, the replacement of long alkyl chains by fluorinated ones provides a better affinity between IL and PS matrix and promotes cell morphology with significantly reduced cell sizes (Figure 5.5).

In conclusion, these first results have highlighted that the ionic liquids can be considered as efficient additives to provide internal surfaces that induces a diffusion restriction and generates a heterogeneous nucleation. The chemical nature of cations plays a key role on the cell morphology generated within polystyrene matrix. The addition of ILs with a CO$_2$-philic moiety, that is, fluorinated chains, is very relevant to enhance foaming by reducing the cell size at lower temperature-pressure conditions. As a consequence, these new foaming agents and their infinite combination of cation–anion offer a new alternative to conventional additives commonly used.

### 5.4.3 Effect of Nanoparticles on Foam Morphology

The use of nanofillers, characterized by their high specific surface and aspect ratio, is another route well known to limit the cell's growth in foam processing (Anastas, 2010, 2012). In the last three decades, the processing of nanocomposite foams has been commonly studied to reduce the nucleation free energy (Tomasko et al., 2003). Well-dispersed nanoparticles act as nucleating sites that can assist in the formation of a nucleation center for the gaseous phase. For example, the use of nanosized particles promotes the creation of nucleation sites at the polymer–particle interface and leads to a higher cell density (Chen, Straff, and Wang, 2001; Park, Behravesh, and Venter, 1997). The quantity and the dispersion state of nanoparticles in the polymer matrix are important parameters that impact the foam morphology (Nam et al., 2002; Strauss and Souza, 2004). The systems most commonly encountered in the literature are based on polymer–clay nanocomposites such as polyolefins, polyurethanes, PVC, polystyrene, polyamide, and biopolymers (Mitsunaga et al., 2003; Ray and Okamoto, 2003b). Thus, Nam et al. (2002) have shown that the cell density of the PP foams increased linearly as a function of the stearyl ammonium modified montmorillonite amount (2, 4, 7.5%wt) (Di et al., 2005). Zhai et al. have obtained a remarkable increase of cell density and a reduction of cell size (0.2–0.3 microns) with increasing nano-silica content (1% to 9%) on polycarbonate nanocomposites (Meissner and Hostettler, 1994). The effect of the nanoparticle distribution on the foam morphology has been also studied in detail (Avalos et al., 2009; Okamoto et al., 2001; Pilla et al., 2010; Zeng et al., 2003). In the case of PS/clay nanocomposites, exfoliated nanocomposites, prepared by *in situ* polymerization, show a better nucleation rate compared to nanocomposites with an intercalated morphology. These results are explained by a greater availability of heterogeneous nucleation sites. Moreover, the nanoparticles have a significant effect on the cell size of the nanocomposite foams. Pilla et al. (2010) have demonstrated that the use of talc in polymer blends of PBAT and PLA led to a decrease in the average cell size. Indeed, as talc increases the crystallinity of the polymer blend, a smaller amount of gas is required for bubble growth in the amorphous phase, which explains the reduction of cell size. Moreover, the addition of nanoparticles has a tendency to increase the viscosity of polymers, which also limits the growth of cells (Okamoto et al., 2001).

The surface functionalization of nanoparticles is an important requirement to control the compatibility between fillers and polymer matrix and as a result the dispersion state of nanofillers. For example, to tune the organophilic character of nanoparticles and promote dispersion within polymers, chemical surface treatments such as cationic exchange, grafting of organosilanes, or the use of copolymers have been extensively practiced (Avalos et al., 2009; Giannelis, 1996; Patel et al., 2007). For example, Ngo et al. (2010) have shown the influence of the chemical nature of the surfactant used for the modification of the layered silicates on polystyrene foam morphologies. Smaller cell sizes and higher nucleation density were determined with a reactive surfactant, 2-methacryloyloxyethyl hexadecyldimethylammonium bromide able to react with styrene monomer unlike to dimethyl benzylammonium chloride (Ngo et al., 2010). In the case of poly(D, L lactid acid)-montmorillonite nanocomposite foams, Tsimpliaraki et al. (2011) have demonstrated that the chain length of the ammonium surfactant is a crucial parameter (C18 instead of C4) to reduce the pore diameter and to increase the pore density. Recently, Goren et al. (2010, 2012) have investigated the influence of silica surface chemistry on pore nucleation in silica/poly(methyl methacrylate) (PMMA) nanocomposites. They have concluded that the surface treatment of nanosilica with fluoroalkanes reduced the critical nucleus radius and decreased the surface free energy. More recently, Yang et al. (2012) have synthesized silica particles modified by a poly(ionic liquid) denoted poly[2-(methacryloyloxy)ethyl]trimethylammonium tetrafluoroborate, which possesses a strong capacity for $CO_2$ adsorption, and they used this modified silica as a nucleation agent in polystyrene foaming. The resulting nanocomposite foam had a smaller cell size and higher cell density compared to PS foams filled with aminosilanes grafted silica (Si-NH2). Other inorganic fillers have also been studied such as graphene or carbon nanotubes (Gedler et al., 2012; Zeng et al., 2010). For example, a comparative study was carried out by Zeng et al. on the influence of chemical treatment of multi-walled carbon nanotubes on the cell morphology and cell density of PMMA foams (Zeng et al., 2010).

In conclusion, nanoparticles are excellent candidates to create microcellular foams because the presence of these inorganic or organic heterogeneities in the polymer matrix promotes nucleation. Nevertheless, the introduction of fillers increases the complexity in the foaming process because in addition to having to control the foaming parameters, it is necessary to carefully select the shape and the size of the nanoparticles as well as that of the chemical nature of the modifier's agents to be placed under the optimum foaming conditions.

### 5.4.4 Influence of Nanoparticles on Foam Properties

Their nanoscale dimensions, responsible for their high specific surface (from 200 to 700 $m^2/g$) and for the significant aspect ratio (from 100 to 1000), confer on them dramatic properties linked to their capacity of confinement within material. Although the majority of studies reported in the literature mainly concern effects of nanoparticles on foam morphology, some studies have shown the benefit of nanoscale fillers on the final properties of polymer nanocomposite foams such as mechanical, thermal, and barrier properties.

Regarding the mechanical properties, it is well known that the addition of a few percent of organically modified layered silicates led to increases of tensile strength; bending strength as well as compressive properties (Han, Zheng, and Lee, 2003; Lee, Lee, and Choi, 2004; Livi et al., 2010). In fact, the addition of 3% of Cloisite 30B in a PVC matrix has allowed an improvement of 250% of the elongation at break and an increase of 26% of bending strength (Lee, Lee, and Choi, 2004). In another study, Han, Zheng, and Lee (2003) studied the influence of different lamellar fillers (talc, montmorillonite) on PS foams. They showed that the nanocomposite foam with an exfoliated morphology led to a weight reduction of 31% and a limited loss of modulus (19%) compared to nonfoamed PS nanocomposite. In terms of compressive properties, it was also demonstrated that the addition of 1 to 5% of carbon nanotubes in PS foams increased the compressive modulus compared to PS in solid state (Shen, Han, and Lee, 2005). By improving the quality of the sample surface, Shen et al. showed significant increases in mechanical properties with an improvement of 81 to 87% of tensile strength and 113 to 124% of impact strength compared to unfoamed PS properties (Chen, Liao, and Chen, 2012). Recently, Chen, Schadler, and Ozisik (2011) investigated the compressive properties of multi-walled carbon nanotube/PMMA nanocomposite foams and demonstrated that the addition of only 1% of MWNT by weight generated an increase of modulus and collapse strength (Figure 5.6).

In addition, the authors developed a constitutive model for predicting the compressive mechanical properties of polymer nanocomposite foams with densities lower than 0.5. Finally, they clearly showed that the morphology controlled the final properties of nanocomposite foams since foams composed of small cells led to better mechanical properties.

In the literature, studies on the thermal stability of foams were also found (Han, Zheng, and Lee, 2003). Han et al. showed an improvement of the fire retardance

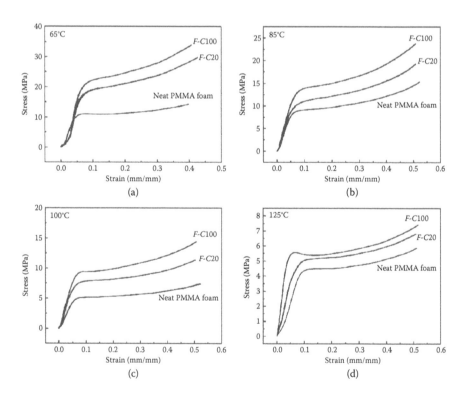

**FIGURE 5.6** Compressive stress–strain curves of nanocomposites and PMMA foamed with ScCO₂ (a) 65°C, (b) 80°C, (c) 100°C, and (d) 125°C. (Reprinted from Chen L. et al., 2011, *Polymer* 52, 2899–2909. Copyright 2013, with permission from Elsevier.)

properties of nanocomposite foams of polystyrene. After burning, PS nanocomposite foams retain their structural integrity while PS foams caused fire spreading (Han, Zheng, and Lee, 2003). More recently, Gedler et al. (2012) studied the thermal decomposition of polycarbonate foams filled with graphene nanoparticles by thermogravimetric analysis (TGA) under nitrogen and air atmosphere. They demonstrated that the barrier effect of graphene nanoplatelets delayed the removal of volatile compounds during degradation and increased the thermal stability of polycarbonate.

In conclusion, these encouraging results highlight the key role of nanoparticles on improving the mechanical and fire retardance properties of nanocomposite foams. Thus, it seems possible to produce lightweight microcellular foams while keeping the final properties of the material in the solid state. However, these early works show the importance of the morphology of the foam on the final properties of polymers. Indeed, a better dispersion of inorganic fillers in the matrix is needed to lead to reduced cell size and high-cell densities. To achieve significantly optimized polymer nanocomposite foams, a lot of studies are needed to find the suitable association between nanoparticles and matrix.

## 5.5  CONCLUSIONS

Currently, polymer nanocomposite foams which represent one part of the lightweight materials commonly used in the plastics industry are the focus of a considerable economic issue. For this reason, significant R&D resources have been devoted to develop a simple and scalable process for commercial production of polymer foams with improved final properties. Moreover, if we add the environmental constraints, supercritical $CO_2$, known for its nontoxicity and its low cost, but also its high solubility in many polymers, has a very promising future. However, despite these advantages, many challenges remain to be overcome before reaching an industrial application. The first challenge is to enhance the development of the foaming process based on injection and extrusion for industry because the existing methods are limited. In fact, the method of batch foaming which is commonly used involves several steps and is not suitable to work with large amounts of polymer. Then, concerning the continuous mode, it is necessary to improve the mathematical models to simulate, predict, and characterize the pressure distribution in different matrices as well as the size distribution of bubbles in order to provide a better understanding of the foaming process. A second challenge is to expand studies on the effect of filler-polymer interactions to control the morphology of the foam in function of the nature of polymer and fillers. Indeed, as we have shown throughout this chapter, the foaming process is very dependent on (i) the nature of the inorganic fillers, (ii) the size, the shape, and the concentration of the nanoparticles, (iii) the surface treatment of particles, and (iv) the compatibility between fillers and polymer matrix which is responsible for the good or poor distribution of fillers. A third challenge is to find a suitable association between the use of additives (ionic liquids, chain extender) to increase the rheological properties of the polymer and the addition of the fillers to limit cell growth while achieving optimized properties of foams.

In conclusion, the development of the *green* process of polymer nanocomposite foams in order to achieve significantly optimized lightweight materials, that is, with good mechanical properties, high surface quality, and excellent thermal and high-dimensional stability, still requires a lot of future work.

## REFERENCES

Alavi, S. H., Khan, M., Bowman, B. J., and Rizvi, S. S. H. 1999. Structural properties of protein-stabilized starch-based supercritical fluid extrudates. *Food Res Int* 31:107–18.

Alavi, S. H., Rizvi, S. S. H., and Harriott, P. 2003a. Process-dynamics of starch-based microcellular foams produced by supercritical fluid extrusion. II: Numerical simulation and experimental evaluation. *Food Res Int* 36:3321–30.

Alavi, S. H., Rizvi, S. S. H., and Harriott, P. 2003b. Process-dynamics of starch-based microcellular foams produced by supercritical fluid extrusion. I: Model development. *Food Res Int* 36: 309–19.

Anastas, P. T. 2010. *Handbook of Green Chemistry: Green Solvents*. NewYork: WileyVCH.

Anastas, P. T. 2012. *Handbook of Green Chemistry: Green Processes*. NewYork: WileyVCH.

Arora, K. A., Lesser, A. J., and McCarthy, T. J. 1998. Preparation and characterization of microcellular polystyrene foams processed in supercritical carbon dioxide. *Macromolecules* 31:4614–4620.

Avalos, F., Ortiz, J. C., Zitzumbo, R., López-Manchado, M. A., Verdejo, R., and Arroyo, M. 2009. Preparation and characterization of phosphonium montmorillonite with enhanced thermal stability. *Appl Clay Sci* 43:27.

Baldwin, D. F., Park, C. B., and Suh, N. P. 1996a. A microcellular processing study of poly(ethylene terephthalate) in the amorphous and semicrystalline states. Part I. Microcell nucleation. *Polymer Engineering and Science* 36:1437–1445.

Baldwin, D. F., Park, C. B., and Suh, N. P. 1996b. A microcellular processing study of poly(ethylene terephthalate) in the amorphous and semicrystalline states. Part II. *Cell Growth and Process Design*, 36:1446–1453.

Beckmann, E. J. and Goel, S. K. 1994. Generation of microcellular polymeric foams using supercritical carbon dioxide. I: Effect of pressure and temperature on nucleation. *Polym Eng Sci* 34:1137–1147.

Błędzki, A. K., Faruk, O., Kirschling, H., Kühn, J., and Jaszkiewicz, A. 2006. Types of foaming agents and technologies of microcellular processing. *Microcellular Polymers and Composites Part I* 51:696–703.

Chen, S. C., Liao, W. H., and Chen, R. D. 2012. Structure and mechanical properties of polystyrene foams made through microcellular injection molding via control mechanisms of gas counter pressure and mold temperature. *International Communications in Heat and Mass Transfer* 39:1125–1131.

Chen, L., Schadler, L. S., and Ozisik, R. 2011. An experimental and theoretical investigation of the compressive properties multi-walled carbon nanotube/Pol(methyl methacrylate) nanocomposites. *Polymer* 52:2899–2909.

Chen, L., Straff, R., and Wang, X. 2001. Effect of filler size on cell nucleation during foaming process. *SPE-ANTEC* 59:1732.

Colton, J. S. and Suh, N. P. 1987. The nucleation of microcellular thermoplastic foam with additives. Part I: Theoretical considerations. *Polym Eng Sci* 27:485–492.

Corre, Y. M., Maazouz, A., Regnier, J., and Duchet-Rumeau, J. 2011a. Batch foaming of chain extended PLA with supercritical $CO_2$: Influence of the rheological properties and the process parameters on the cellular structure. *Journal of Supercritical Fluids* 58: 177–188.

Corre, Y. M., Maazouz, A., Regnier, J., and Duchet-Rumeau, J. 2011b. Melt strengthening of poly(lactic acid) through reactive extrusion with epoxy-functionalized chains. *Rheologica Acta* 50:1–17.

DeSimone, J. M., Maury, E. E., Menceloglu, Y. Z., McClain, J. B., Romack, T. J., and Combes, J. R. 1994. Dispersion polymerizations in supercritical carbon dioxide. *Science* 265:356.

Di, Y., Iannace, S., Di Maio, E., and Nicolais, L. 2005. Poly(lactic acid)/organoclay nanocomposites: Thermal, rheological properties and foam processing. *J Polym Sci, Part B: Polym Phys* 43:689.

Doroudiani, S., Park, C. B., and Kortschot, M. T. 1996. Effect of the crystallinity and morphology on the microcellular foam structure of semicrystalline polymers. *Polymer Engineering and Science* 36:2645–2662.

Doroudiani, S., Park, C. B., and Kortschot, M. T. 1998. Processing and characterization of microcellular foamed high-density polythylene/isotactic polypropylene blends. *Polym Eng Sci* 38:1205–1215.

Gedler, G., Antunes, M., Realinho, V., and Velasco, J. I. 2012. Thermal stability of polycarbonate-graphene nanocomposite foams. *Polymer Degradation and Stability* 97:1297–1304.

Gendron, R. and Daigneault, L. E. 2000. S. T. Lee (Ed.), *Foams Extrusion: Principle and Practice*, Lancaster: Technomic Publ.

Giannelis, E. P. 1996. Polymer Layered Silicate Nanocomposites. *Adv Mater* 8:29.

Goren, K., Chen, L., Schadler, L. S., and Ozisik, R. 2010. Influence of nanoparticle surface chemistry and size on supercritical carbon dioxide processed nanocomposite foam morphology. *J of Supercritical Fluids* 51:420–427.

Goren, K., Okan, O. B., Chen, L., Schadler, L. S., and Ozisik, R. 2012. Supercritical carbon dioxide assisted dispersion and distribution of silica nanoparticles in polymers. *J of Supercritical Fluids* 67:108–113.

Han, X., Zheng, C., and Lee, L. J. 2002. Processing and cell structure of nano-clay modified microcellular foams. *ANTEC* 2002.

Han, X., Zheng, C., and Lee, L. J. 2003. Extrusion of polystyrene nanocomposite foams with supercritical CO$_2$. *Polym Eng Sci* 43:1261–1275.

Jacobsen, K. and Pierick, D. 2000. Microcellular foam molding: Advantages and applications *ANTEC* 2000 2:1929–1233.

Jacobsen, K. and Pierick, D. 2001. Injection molding innovation: The microcellular foam process. *Plast Eng* 57:46–51.

Jiang, X. L., Liu, T., Xu, Z. M., Zhao, L., Hu, G. H., and Yuan, W. K. 2009. Effects of crystal structure on the foaming of isotactic polypropylene using supercritical carbon dioxide as a foaming agent. *Journal of Supercritical Fluids* 48:167–175.

Jin, W., Xingguo, C., Mingjun, Y., and Jiasong, H. 2001. An investigation on the microcellular structure of polystyrene/LCP blends prepared by using supercritical carbon dioxide. *Polymer* 42:8265–8275.

Kwag, C., Manke, C. W., and Gulari, E. 1999. Rheology of molten polystyrene with dissolved supercritical and near-critical gases. *Journal of Polymer Science Part B: Polymer Physics* 37:2771–2781.

Lee, S. T., Kareko, L., and Jung, J. 2008. Study of thermoplastic PLA foam extrusion. *J Cell Plast* 44:293–305.

Lee, M., Lee, B. K., and Choi, K. D. 2004. Foam compositions of polyvinyl chloride nanocomposites, *PCT Int Appl* 2004074357.

Lian, Z., Epstein, S. A., Blenk, C. W., and Shine, A. D. 2006. Carbon dioxide-induced melting point depression of biodegradable semicrystalline polymers. *Journal of Supercritical Fluids* 39:107–117.

Lim, L. T., Auras, R., and Rubino, M. Processing technologies for poly(lactid acid). 2008. *Progress in Polymer Science* 33:820–852.

Livi, S., Duchet-Rumeau, J., Pham, T. N., and Gérard, J. F. 2010. A comparative study on different ionic liquids used as surfactants: Effect on thermal and mechanical properties of high density polyethylene nanocomposites. *Journal of Colloid and Interface Science* 349:424–433.

Livi, S., Gérard, J. F., and Duchet-Rumeau, J. 2011a. Ionic liquids: Structuration agents in a fluorinated matrix. *Chem Commun* 47:3589–3591.

Livi, S., Gérard, J. F., and Duchet-Rumeau, J. 2011b. Nanostructuration of ionic liquids in a fluorinated matrix. *Polymer* 52:1523–1531.

Livi, S., Pham, T. N., Gérard, J. F., and Duchet-Rumeau, J. 2013. Supercritical $CO_2$-Ionic Liquids: Green combination for preparing foams. Submitted.

Mascia, L., Re, G. D., Ponti, P. P., Bologna, S., Giacomo, G. D., and Haworth, B. 2006. Crystallization effects on autoclave foaming of polycarbonate using supercritical carbon dioxide. *Advances in Polymer Technology* 25:225–235.

Matuana, L. M. 2008. Solid state microcellular foamed poly(lactic acid): Morphology and property characterization. *Bioresource Technology* 99:3643–3650.

Mawson, S., Johnston, K. P., Combes, J. R., and DeSimone, J. M. 1995. Formation of poly(1,1,2,2-tetrahydroperfluorodecyl acrylate) submicron fibers and particles from supercritical carbon dioxide solutions. *Macromolecules* 28:3182.

McHugh, M. A., Garach-Domech, A., Park, I. H., Li, D., Barbu, E., Graham, P., and Tsibouklis, J. 2002. Impact of fluorination and side-chain length on poly(methyl propenoxyalkylsiloxane) and poly(alkyl methacrylate) solubility in supercritical carbon dioxide. *Macromolecules* 35:6479.

Meissner, J. and Hostettler, J. 1994. A new elongational rheometer for polymer melts and other highly viscoelastic liquids. *Rheologica Acta* 33:1–21.

Michaeli, W. and Heinz, R. 2000. Foam extrusion of thermoplastics polyurethanes (TPU) using $CO_2$ as a blowing agent. *Macromol Mater Eng* 35:284–285.

Mihai, M., Huneault, M. A., Favis, B. D., and Li, H. 2007. Foaming of PLA/thermoplastic starch blends. *Macromol Biosci* 7:907–920.

Mihai, M., Huneault, M. A., and Favis, B. D. 2010. Rheology and extrusion foaming of chain-branched poly(lactic acid). *Polymer Engineering and Science* 50:629–642.

Mitsunaga, M., Ito, Y., Ray, S. S., Okamoto, M., and Hironako, K. 2003. Intercalated polycarbonate/clay nanocomposites: Nanostructure control and foam processing. *Macromol Mater Eng* 288:543.

Naguib, H. E., Park, C. B., and Song, S. W. 2005. Effect of supercritical gas on crystallization of linear and branched polypropylene resins with foaming additives. *Industrial and Engineering Chemistry Research* 44:6685–6691.

Nalawade, S. P., Picchioni, F., and Janssen, L. P. B. M. 2006. Supercritical carbon dioxide as a green solvent for processing polymer melts: Processing aspects and applications. *Progress in Polymer Science* 31:19–43.

Nam, P. H., Maiti, P., Okamoto, M., Kotaka, T., Nakayama, T., Takada, M. et al. 2002. Foam processing and cellular structure of polypropylene/clay nanocomposites. *Polym Eng Sci* 42:1907.

Nam, G. J., Yoo, J. H., and Lee, J. W. 2005. Effect of long-chain branches of polypropylene on rheological properties and foam-extrusion performances. *Journal of Applied Polymer Science* 96:1793–1800.

Ngo, T. T. V., Duchet-Rumeau, J., Whittaker, A. K., and Gerard, J. F. 2010. Processing of nanocomposite foams in super critical carbon dioxide. Part I: Effect of surfactant. *Polymer* 51:3436–3444.

Okamoto, M., Nam, P. H., Maiti, P., Kotaka, T., Nakayama, T., and Takada, M. et al. 2001. Biaxial flow-induced alignment of silicate layers in polypropylene/clay nanocomposite foam. *Nano Lett* 1:503.

Park, C. B., Baldwin, D. F., and Suh, N. P. 1995. Effect of the pressure drop on cell nucleation in continuous processing of microcellular polymers. *Pol Eng Sci* 35:432–40.

Park, C. B., Behravesh, A. H., and Venter, R. D. 1997. A strategy for the suppression of cell coalescence in the extrusion of microcellular high-impact polystyrene foams. *ACS Symp Ser* (Polymeric Foams) 669:115.

Park, C. B. and Suh, N. P. 1993. SPE *ANTEC Tech. Papers* 39:1818.

Patel, H. A., Somani, R. S., Bajaj, H. C., and Jasra, R. V. 2007. Phosphonium salt intercalated montmorillonites. *Appl Clay Sci* 35:194.

Pilla, S., Kim, S. G., Auer, G. K., Gong, S., and Park, C. B. 2009. Microcellular extrusion-foaming of polylactide with chain-extender. *Polymer Engineering and Science* 49:1653–1660.

Pilla, S., Kim, S. G., Auer, G. K., Gong, S., and Park, C. B. 2010. Microcellular extrusion foaming of poly(lactide)/poly(butylene adipate-co-terephtalate) blends. *Materials Science and Engineering* C 30:255–262.

Ray, S. S. and Okamoto, M. 2003a. Polymer–layered silicate nanocomposite: A review from preparation to processing. *Prog Polym Sci* 28:1539–1641.

Ray, S. S. and Okamoto, M. 2003b. New polylactide/layered silicate nanocomposites. Part 6. Melt rheology and foam processing. *Macromol Mater Eng* 288:936.

Reignier, J., Gendron, R., and Champagne, M. F. 2007. Autoclave foaming of poly({varepsilon}-caprolactone) using carbon dioxide: Impact of crystallization on cell structure. *Journal of Cellular Plastics* 43:459–489.

Reverchon, E. and Cardea, S. 2007. Production of controlled polymeric foams by supercritical $CO_2$. *Journal of Supercritical Fluids* 40:144–152.

Sauceau, M., Fages, J., Common, A., Nikitine, C., and Rodier, E. 2011. New challenges in polymer foaming: A review of extrusion processes assisted by supercritical carbon dioxide. *Prog Polym Sci* 36:749–766.

Sauthof, R. 2003. Physical foaming with Ergocell. Blowing agents and foaming processes. 2003. Munich: *Rapra Conference Proceedings*.

Shen, J., Han, X., and Lee, L. J. 2005. Nucleation and reinforcement of carbon nanofibers on polystyrene nanocomposite foam. 63rd ed. Annual Technical Conference, *Society of Plastics Engineers* p. 1896. ASAP.

Soares, B. G., Livi, S., Gérard, J. F., and Duchet-Rumeau, J. 2012. Preparation of epoxy/MCDEA networks modified with ionic liquids. *Polymer* 53:60–66.

Spitael, M. C. W. P. 2004. Strain hardening in polypropylenes and its role in extrusion foaming. *Polymer Engineering and Science* 44:2090–2100.

Stafford, C. M., Russell, T. P., and McCarthy, T. J. 1999. Expansion of polystyrene using supercritical carbon dioxide: Effects of molecular weight, polydispersity, and low molecular weight. *Macromolecules* 32:7610–7616.

Stange, J. and Munstedt, H. 2006. Effect of long-chain branching on the foaming of polypropylene with azodicarbonamide. *Journal of Cellular Plastics* 42:445–467.

Strauss, W. D. and Souza, N. A. 2004. Supercritical $CO_2$ processed polystyrene nanocomposite foams. *J Cell Plast* 40:229.

Sulzer. http://www.sulzerchemtech.com/en/desktopdefault.aspx/tabid-568/733_read-8650/.

Tai, H., Upton, C. E., White, L. J., Pini, R., Storti, G., and Mazzotti, M. 2010. Studies on the interactions of $CO_2$ with biodegradable poly(dl-lactic acid) and poly(lactic acid-co-glycolicacid) copolymers using high pressure ATR-IR and high pressure rheology. *Polymer* 51:1425–1431.

Takada, M., Hasegawa, S., and Ohshima, M. 2004. Crystallization kinetics of poly(l-lactide) in contact with pressurized $CO_2$. *Polymer Engineering and Science* 44:186–196.

Tatibouet, J. and Gendron, R. 2005. Heterogeneous nucleation in foams as assessed by in-line ultrasonic measurements. *Journal of Cellular Plastics* 41:57–72.

Tomasko, D. L., Han, X., Liu, D., and Gao, W. 2003. *Current Opinion in Solid State and Materials Science* 7:407–412.

Tsimpliaraki, A., Tsivintzelis, I., Marras, S. I., Zuburtikudis, I., and Panayiotou, C. 2011. The effect of surface chemistry and nanoclay loading on the microcellular structure of porous poly(D,L lactid acid) nanocomposites. *Journal of Supercritical Fluids* 57:278–287.

Tsivintzelis, I., Angelopoulou, A. G., and Panayiotou, C. 2007. Foaming of polymers with supercritical $CO_2$: An experimental and theoretical study. *Polymer* 48:5928–5939.

Wang, J., Zhu, W., Zhang, H., and Park, C. B. 2012. Continuous processing of low-density, microcellular poly(lactic acid) foams with controlled cell morphology and crystallinity. *Chemical Engineering Science* 75:390–399.

Wissinger, R. G. and Paulaitis, M. E. 1991. Glass transitions in polymer/$CO_2$ mixtures at elevated pressures. *Journal of Polymer Science Part B: Polymer Physics* 29:631–633.

Wolff, F., Zirkel, L., Betzold, S., Jakob, M., Maier, V., Nachtrab, F., Ceron, N., Nicolat, B., Fey, T., and Munstedt, H. 2011. Using supercritical carbon dioxide for physical foaming of advanced polymer materials. *Intern Polymer Processing* XXVI 4 (2011).

Yang, J., Sang, Y., Chen, F., Fei, Z., and Zhong, M. 2012. Synthesis of silica particles grafted with poly(ionic liquid) and their nucleation effect on microcellular foaming of polystyrene using supercritical carbon dioxide. *J of Supercritical Fluids* 62:197–203.

Zeng, C., Han, X., Lee, L. J., Koelling, K. W., and Tomasko, D. L. 2003. Polymer–clay nanocomposite foams prepared using carbon dioxide. *Adv Mater* (Weinheim, Germany) 15:1743.

Zeng, C., Hossieny, N., Zhang, C., and Wang, B. 2010. Synthesis and processing of PMMA carbon nanotube nanocomposite foams. *Polymer* 51:655–664.

Zhai, W., Yu, J., Ma, W., and He, J. 2006. Influence of long-chain branching on the crystallization and melting behavior of polycarbonates in supercritical $CO_2$. *Macromolecules* 40:73–80.

Zhai, W., Yu, J., Wu, L., Ma, W., and He, J. 2006. Heterogeneous nucleation uniformizing cell size distribution in microcellular nanocomposites foams. *Polymer* 47:7580–7589.

Zhou, Z. F., Huang, G. Q., Xu, W. B., and Ren, F. M. 2007. Chain extension and branching of poly(l-lactic acid) produced by reaction with a DGEBA-based epoxy resin. *Express Polymer Letters* 1:734–739.

# 6 Hybrid Polyurethane Nanocomposite Foams

*Marcelo Antunes*

## CONTENTS

## 6.1 INTRODUCTION

Polyurethane (PU) foams are a family of cellular materials that are used extensively in almost every ambit of application due to their extended range of physical and chemical properties, which depend on the formulation used and processing. Though a great deal of attention has been given to overcoming the relatively low thermal stability and especially poor mechanical properties of polyurethane foams, which have mainly been achieved through the years by the initial replacement of polyester polyols by polyether polyols and later by incorporation of foam catalysts and stabilizers, enabling a vast implementation of one-step polyurethane foaming technologies and thus a general boost in polyurethane foaming, there are still limitations in terms of properties that considerably limit their range of application. For instance, owing to their extremely reduced thermal conductivity, low-density rigid PU foams are used extensively as thermal insulators in the building sector. Nevertheless, the extremely low densities required to achieve such a level of thermal insulation result in significant decreases in terms of mechanical strength, inhibiting the use of these foams in many structural applications.

The inclusion of reinforcements, especially inorganic ones such as glass or carbon fiber, has always been seen as a strategy to enhance the mechanical properties of polymers, resulting in hybrid materials that are known as polymer composites. The concept of polymer composite has recently been applied to polymer foams, and particularly to polyurethane foams, with the main objective of extending the applicability of said foams by globally improving their mechanical response. Even more recently, researchers have tried to further extend this concept to the field of nanoscience by incorporating nanosized reinforcements into polyurethane foams, as it has been shown that nanosized particles may, under ideal conditions, have an extremely high interaction area with polymer molecules, effectively reinforcing the foam even at low added concentrations. Some extra benefits could also result from creating hybrid PU nanocomposite foams, as well-dispersed nanoparticles could promote the formation of finer and more uniform foam cellular structures, favoring an increase in mechanical strength and even in thermal insulation. In any case, the addition of relatively low concentrations of nanosized particles to PU foam formulations, both flexible and rigid-like, has been proven to be a viable strategy to extend the range of possible applications of polyurethane foams.

In addition to the common objective of improving the mechanical performance of PU foams, researchers have also focused on the possibility of attaining other characteristics for PU foams by introducing functional nanoparticles, that is, nanometric-sized particles that alongside their high mechanical properties may have other interesting features, such as high transport properties or inherent flame retardancy. Some of the most used functional nanoparticles include organically modified nanoclays, carbon-based nanofillers, particularly carbon nanotubes, nanofibers, and more recently graphene, nanosilica, and biobased nanofillers. The inclusion of these functional nanoparticles could result in the development of hybrid PU foams with a vast array of properties, thus opening up brand new possible applications.

This chapter considers the development and main properties of hybrid polyurethane nanocomposite foams, flexible as well as rigid, focusing on the influence of processing and incorporation of various types of nanometric-sized fillers in the structure, and mechanical properties, transport properties, and other significant properties of the resulting foams. Special interest has been given to new developments that have recently emerged for hybrid polyurethane nanocomposite foams in the most varied fields, particularly the influence of incorporating novel nanohybrids based on the combination of different nanoparticles into polyurethane foams and the resulting properties, the use of new hybrid polyurethane foams as scaffolds for tissue engineering applications, and the concept of creating rigid polyurethane foams with improved properties by combining multiscalar reinforcements.

## 6.2 POLYURETHANE FOAMS: BASICS AND APPLICATIONS

As indicated by their name, polyurethanes (PUs), also referred to as *urethanes*, are a polymer family characterized by the presence of the urethane group (-NH-CO-O-), which is formed by the addition reaction between isocyanate (-N=C=O) and hydroxyl groups (-OH). Polyurethane foams, commonly known as *PU foams, PUR foams*, or simply *PUF*, occupy the largest polymer foam market share in terms of consumption, their use being vastly extended to a great variety of applications. Depending on its particular formulation and processing, a vast array of PU foams may be obtained in terms of cellular structure (closed cell, interconnected, or fully open celled), and global mechanical characteristics, from soft and flexible to rigid-like (Eaves, 2004).

Though PU foam formulations tend to be quite complex and include several components such as surfactant(s) or catalyzer(s), usually added in small quantities (typically less than 1 to 2%), or even other components added in higher amounts, such as flame retardants (consult Ashida, 2007; Backus, 2004; Eaves, 2004; Herrington, Broos, and Knaub, 2004, for further information), PU foaming can be seen as the reaction of two basic components, an isocyanate and a polyol or mix of polyols, both liquid at room temperature. Foaming is possible due to the addition of a physical blowing agent or more usually the *in situ* generation of a blowing agent ($CO_2$) simultaneous to polyurethane's polymerization. The formation of $CO_2$ follows a two-step reaction that starts out by the reaction of isocyanate with water (formulated in a given proportion with the polyol[s] component[s]) to give unstable carbamic acid and the later strongly exothermal decomposition of this acid (see Figure 6.1).

Depending on the choice of isocyanate and polyol starting materials, R and R' groups may be tailored to contain isocyanate and isocyanate-reactive groups, respectively, enabling the formation of polymers with a cross-linked soft-segment (given by the polyol portions)–hard-segment (isocyanate segments) network, which have

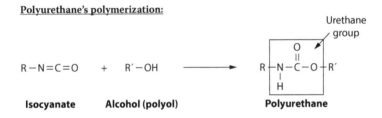

**FIGURE 6.1**  Basic polymerization and gas-producing reactions occurring during PU foaming.

a direct effect in guaranteeing a proper stabilization of the *in situ* generated foam. Further reaction of the amine generated in the gas-producing reaction with excess isocyanate gives a disubstituted urea, which can further react with isocyanate to create biuret linkages between polymer molecules (cross-linking) (Herrington, Broos, and Knaub, 2004).

As can be seen, one of the main ways of controlling the final characteristics of PU foams consists of a proper material selection and formulation, that is, in a proper choice and proportion of both isocyanate and polyol(s) components.

In the case of the polyol, the first industrially produced PU foams used a polyester-type polyol. Nevertheless, PU foams obtained by the reaction of said polyol with an aromatic polyisocyanate presented many problems in terms of foam stability and long-term use. Also, due to the high exothermicity resulting from the primary reaction between polyol and isocyanate, one-step PU foaming processes (also known as *one-shot* PU foam technologies) that used polyester polyols normally resulted in foam scorching or even burning, especially in the case of high-thickness products such as foamed blocks. In order to avoid these problems, a two-step process consisting in the initial formation of an isocyanate-terminated urethane prepolymer and later reaction of this prepolymer with water to give the blowing agent was implemented, enabling guarantee of a stable foam growth without later collapse (Ashida, 2007).

Improved performance was especially attained by replacing polyester polyols with polyether polyols, with the resulting PU foams presenting a lower sensitivity to hydrolysis and higher durability. The combination at the beginning of the 1960s of polyether polyols, high-purity isocyanates, and new catalysts (such as 1,4-diazabicyclo [2,2,2] octane or DABCO) (Farkas et al., 1959) and surfactants (mainly based on polysiloxane-polyoxyalkylene block copolymers) made viable the industrial use of *one-shot* PU foam technologies, leading to a big growth in PU foaming industry. As a matter of fact, more than 90% of all flexible PU foams fabricated nowadays are produced from polyether polyols. In addition to polyether and increasingly disused polyester polyols, copolymer polyols obtained by free radical grafting of styrene and acrylonitrile (SAN copolymer) to polypropylene glycol (PPG) are also being used in specific foam applications. The main types of polyols used in PU foaming are summarized in Figure 6.2.

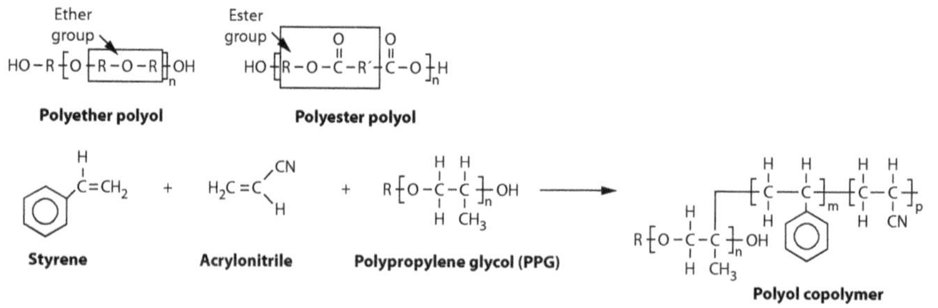

**FIGURE 6.2** Main types of polyols used in PU foaming: polyether polyol, polyester polyol, and polyol copolymer.

In terms of isocyanate selection, the types presently used at the industrial level contain at least two reactive isocyanate groups per molecule. While toluene diisocyanate (TDI) is the most used isocyanate in the production of flexible PU foams, methylene diphenyl diisocyanate (MDI) is the most extended one in the preparation of semiflexible and rigid PU foams (see Figure 6.3).

As can be seen, polyurethane foams are basically classified according to their global mechanical behavior in flexible and rigid foams. A third type with intermediate mechanical properties is sometimes considered: semirigid or semiflexible PU foams.

Though the final flexibility of the produced foam depends on the formulation used, particularly in terms of isocyanate and polyol(s) selection and their respective proportions, as well as the amount of blowing agent and foaming process (please consult Section 6.3 for further details on blowing agent considerations and especially polyurethane foaming technologies), generally speaking and in contrast to the open-cell structure of flexible PU foams, rigid foams usually present a closed-cell structure, which accounts, in great part, for their higher strength-to-weight ratio as well as improved thermal insulation.

A classification of PU foams in rigid, semirigid, and flexible foams is presented in Table 6.1. As can be seen, rigid PU foams are characterized by their extremely high

**Semi-flexible and rigid PU foams:** **Flexible PU foams:**

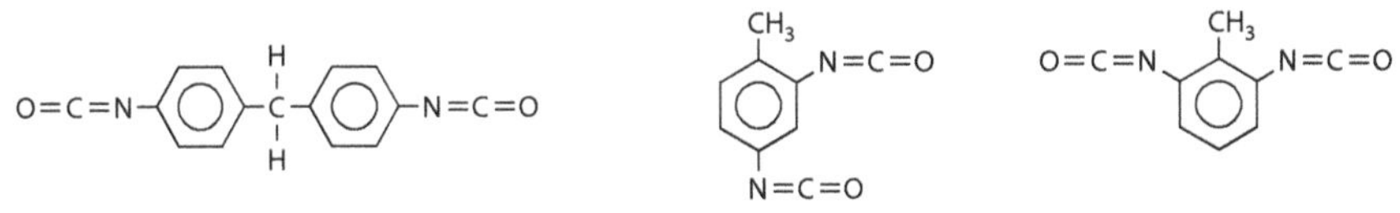


**Methylene diphenyl diisocyanate (MDI)** 2,4-Toluene diisocyanate 2,6 - Toluene diisocyanate

**Isomers of toluene diisocyanate (TDI)**

**FIGURE 6.3** Most common isocyanates used in the production of flexible and semiflexible/rigid PU foams.

## TABLE 6.1
## Classification of PU Foams in Rigid, Semirigid, and Flexible Foams

| Characteristics of the Polyol | Rigid Foam | Semirigid Foam | Flexible Foam |
|---|---|---|---|
| OH number | 350–560 | 100–200 | 5.6–70 |
| OH equivalent number | 160–100 | 560–280 | 10000–800 |
| Functionality | 3.0–8.0 | 3.0–3.5 | 2.0–3.1 |
| Elastic modulus at 23°C (MPa) | > 700 | 70–700 | < 70 |

*Source:* Adapted from Ashida, K., 1994, In *Handbook of Plastic Foams*, ed. A. H. Landrock, Park Ridge, NJ: Noyes Publications.

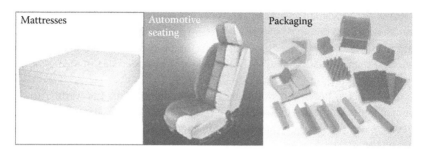

**FIGURE 6.4 (See color insert.)**   Some of the possible applications of flexible PU foams.

elastic modulus, especially when compared with flexible PU foams, a direct result of their particular closed-cell structure and formulation. For instance, polyols used in rigid PU formulation present a much higher functionality and thus reactivity than those used in flexible PU foams, in part explaining their final closed-cell structure and improved mechanical properties.

In terms of applications, flexible PU foams are mostly used in cushioning in fields such as furnishing (furniture, mattress industry, etc.), transportation (automotive seating, vibration damping components, filters, headliners, etc.), protective packaging, and others (electronics, sports, footwear, leisure products, clothing …) (see Figure 6.4), though highly technical foams have also been used in specific applications (Housel, 2004). Due to their high versatility in terms of load-bearing capabilities, densities, and other physical properties, a direct result of a proper selection of the formulation and manufacturing process, flexible PU foams are clearly the most extended type of polymer foam used in protection and comfort applications.

Most of the applications and consumption growth of low-density semirigid and rigid PU foams come from their excellent low- and high-temperature thermal insulation characteristics and the increasing growing interest in reducing energy consumption. For these reasons, low-density rigid PU foams are mostly used in thermal insulation applications and in construction as building materials. Owing to their extremely low thermal conductivity, these foams are employed in thin-wall designs, as for instance in the thermal insulation of building roofs, walls, and windows, doors of trucks, railcars, or containers.

As a direct result of their good strength-to-weight ratio, medium-high-density rigid PU foams are used extensively as structural elements in the construction and automotive sectors (Eaves, 2004), where for instance they are employed as lightweight cores in load-bearing sandwich-like parts, in some cases in combination with metal or wood outer layers. Some other important applications include thermal insulation in cold storing or transport components, as well as their use as impact energy absorbers for car or train protection. Due to their good sound damping properties, open-cell or highly cross-linked rigid PU foams are used in acoustic applications. Some of the main applications of rigid PU foams are presented in Figure 6.5.

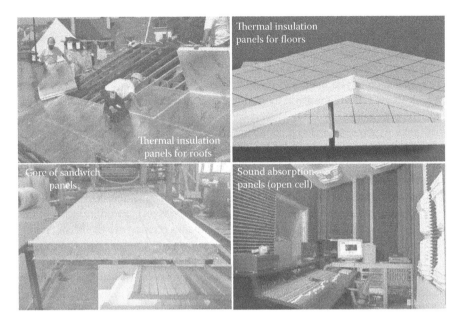

Thermal insulation panels for floors

Thermal insulation panels for roofs

Core of sandwich panels

Sound absorption panels (open cell)

**FIGURE 6.5 (See color insert.)** Some of the possible applications of rigid PU foams.

## 6.3 POLYURETHANE FOAMING TECHNOLOGIES

Polyurethane foam preparation is industrially classified according to the final characteristics of the foam in flexible and rigid PU foaming technologies. In both cases, the technology used must take into account the particularities of PU foams, which are basically prepared by the chemical reaction of two liquids, a polyol, or more commonly a mix of polyols formulated with water or liquid physical blowing or coblowing agents and other processing additives, and an isocyanate (Herrington, Broos, and Knaub, 2004). These components are mixed in the proportions required by each specific formulation, particularly in terms of final mechanical behavior (flexible/rigid), density, and cellular structure (open cell/closed cell), among others. Manual weighing of the reactants or metering feeding pumps that transport the components to a mixing head may be used. After feeding, the components are mixed and cell nucleation is favored, as they chemically react and liberate the blowing agent as polymerization takes place. A cream is formed and there is an initial rise of the foam, followed by its expansion and final stabilization and gellation. A final curing step is needed in order to fully cure the PU matrix and give the required dimensional stability to the foam (Herrington, Broos, and Knaub, 2004).

### 6.3.1 FLEXIBLE POLYURETHANE FOAMS

Flexible PU foams, also known as FPF, are made primarily from the chemical reaction of a polyol or mix of polyols formulated with a given amount of water and other additives such as catalysts or surfactants, and an isocyanate. Depending on the type of polyol, flexible PU foams may be classified into polyether polyol, polyester polyol, or polyol copolymer foams (Ashida, 2007). Nevertheless, the most common classification of FPF is according to their production foaming process, being divided in slabstock and molded foams. The main difference between both processes lies in the fact that slabstock PU foams are prepared by free-foaming in an open mold, while, as clearly indicated by its name, molded PU foams are prepared by pouring the reacting mixture into a mold, which is then closed allowing the foam to grow until complete filling. As flexible PU foams prepared by the slabstock process are prepared by free-foaming, they usually require a final cutting step of the upper part (due to geometrical constraints and nonuniform foam expansion), which in some cases may represent an important amount of disposable material (scrap). Despite all these limitations, more than 50% of all FPF is produced using the slabstock process, owing to its easiness and low cost. Basically, the FPF slabstock foaming process consists of the continuous pouring of a liquid containing all of the above-mentioned required foaming components on a moving conveyor, which results in the continuous formation of a free-rising foam that after cutting is commonly known as *block* or *bun foam*. A typical slabstock equipment configuration includes storage tanks for each of the reacting components, weigh tanks in charge of feeding the components in the exact required proportions to a mixing vessel or head, and an open mold or box with the desired dimensions, where the reacting mixture is poured and the foam produced. In some specific cases some of the ingredients are preblended prior to their feeding to the mixing vessel. Some examples of component preblending include a blend of water and catalyst, polyol and catalyst, or polyol and physical blowing agent (Ashida, 2007). Though the so-called inclined conveyor slabstock process has been widely used since the beginning of the FPF industry (see Figure 6.6), it presents

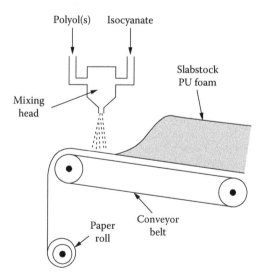

**FIGURE 6.6** Inclined conveyor slabstock PU foam production line. (Adapted from Hodlur, R. M. and Rabinal, M. K., 2012, *Solid St. Phys. Pts. 1 and 2 AIP Conference Proceedings* 1447, 1279–1280.)

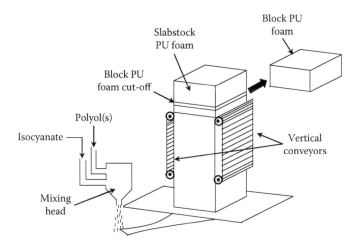

**FIGURE 6.7** Basic scheme of the Vertifoam vertical foaming process to obtain flexible PU foams. (Adapted from Ashida, K., 2007, In *Polyurethane and Related Foams. Chemistry and Technology*, ed. Ashida K., 67–82. New York: CRC Press/Taylor & Francis.)

some geometrical limitations that result in the generation of an important amount of scrap and thus reduced production. Further processing developments considered the way to obtain continuous flexible PU foams with uniform rectangular cross sections. Particularly, due to the particular disposition of the conveyor, a vertical foaming process called *Vertifoam* (Ashida, 2007) has enabled significant increase in the volume of produced FPF while keeping plant space (see Figure 6.7).

Though for many years the most used blowing agent in FPF slabstock foams had been a mix of water and CFC-11, the progressive ban of chlorofluorocarbons (CFCs) and HCFCs after the Montreal Protocol on Substances that Deplete the Ozone Layer of 1987 led to the replacement of CFC-11 by other alternative blowing agents, mainly HFCs such as HFC-245fa, HFC-365mfc, or blends. Nevertheless, the particular high cost of these HFCs has limited their use to specific applications and the general use of water as the only blowing agent. Other blowing agents have also been used, especially carbon dioxide.

As the maximum amount of water is limited to around 4.8 parts per hundred of polyol (php) due to problems related to the dissipation of the heat generated by the strongly exothermic reaction of isocyanate with water, industrially-produced FPF density is limited to about 20 kg/m³, though the use of HFC coblowing agents may slightly reduce this density.

Due to their particular combination of density, and mechanical and geometrical characteristics, slabstock flexible PU foams are widely used in transportation and cushioning applications.

As a direct consequence of equipment requirements such as venting or automatic opening and closing of the mold, or formulation aspects such as faster reacting times, molded flexible PU foam production is quite more complex than that of slabstock foams and considerably more expensive. Depending on the type of curing, two major mold-foaming processes are industrially available: the hot-cure and the cold-cure.

Cold-cure differentiates from hot-cure molding by the fact that low mold temperatures are used, typically between 60 and 70°C, and by the use of higher-reactive components, considerably reducing oven curing time. Postcure in a separate oven or curing at room temperature is often required after demolding the part. High-resilience (HR) PU foams, characterized by their high elastic properties and higher resiliency when compared to conventional flexible PU foams (sag factor typically between 2.4 and 3.0), are produced using the cold-cure foaming process.

Due to their particular sandwich-like structure formed by high density or even solid skins and a low-density foamed core (PU foams with this type of structure are also known as integral foams), molded flexible PU foams are commonly used in automobile interior parts, furniture cushions, or mattresses.

### 6.3.2 Rigid Polyurethane Foams

Rigid PU foams mainly differentiate from flexible PU foams by the fact that their cellular structure is almost a fully closed cell, resulting in foams with low flexibility and high-load-bearing capacity, as well as high thermal insulation.

Contrary to flexible PU foams, which are mainly produced using water as the only or main blowing agent, rigid PU foams use a liquid physical blowing agent or alternatively a pressurized gas added to one of the components. This last one generates the foamed structure when the mixed components are suddenly depressurized. The arrival of the so-called third-generation blowing agents, which came to substitute ozone depletion potential (ODP) CFC and HCFC physical blowing agents, has resulted in the biggest technological development in rigid PU foam processing to date. These third-generation blowing agents are mostly HFCs, especially HFC-245fa, HFC-365mfc, and $C_5$-hydrocarbons such as pentane or cyclopentane, though halogen-free azeotropes and especially carbon dioxide have also been used (Ashida, 2007).

Though strictly speaking, rigid PU foams are those derived from the reaction of an isocyanate (usually methylene diphenyl diisocyanate, MDI) with a short chain polyol or mix of polyols, conventional polyisocyanurate (PIR) foams have also been included as a type of rigid PU foam. PIR production is possible by introducing excess MDI into the mix, as MDI reacts with itself to form cyclic trimeric isocyanurate groups. Some of the advantages of PIR foams when compared to conventional rigid PU foams include higher thermal stability and improved fire retardancy (Ashida, 2007).

Rigid PU foams are basically produced by continuous or discontinuous processes in the form of blocks and sheets (lamination technology for panel production), poured in place, molded, or sprayed. Generally speaking, the equipment required to prepare rigid PU foams include individual component storage tanks, metering pumps to feed the components to a continuous mixer or mixing head, and a final forming device (see diagram presented in Figure 6.8) (Backus, 2004). With the replacement of CFC blowing agents by highly flammable hydrocarbons, rigid PU foam processing equipments had to be adapted in order to avoid possible flammable vapor escape.

As mentioned at the end of Section 6.2, the direct consequence of their particular closed-cell structure and versatile combination with other materials, enabling lamination or the preparation of sandwich panels, rigid PU foams are extensively

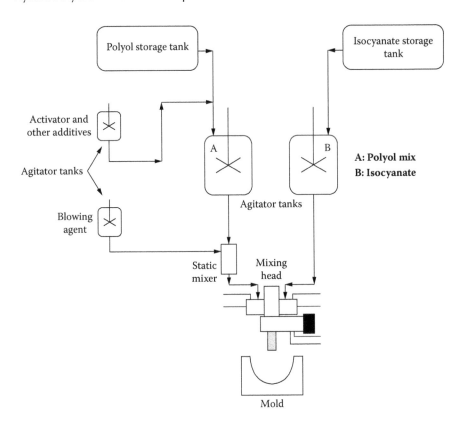

**FIGURE 6.8**  Basic diagram of the equipment required to prepare rigid PU foams.

used as thermal insulating structural elements in refrigerators, refrigerated trucks and containers, building and construction, refrigerated consumer goods, and so forth.

Increasing interest has been given to the possibility of tailoring the properties of both flexible and rigid PU foams by introducing secondary components, mainly inorganic in nature. The combination of incorporating secondary components into a PU matrix and foaming results in the formation of hybrid PU foams that, depending on the dimensions of said component, are also known as *hybrid PU composite* or *PU nanocomposite* foams. Usually, the incorporation of these secondary materials addresses the need to counteract the loss in mechanical properties resulting from foaming, though functional (nano)particles have recently been considered as a possible strategy to regulate or even introduce specific characteristics to PU foams, such as electrical conduction (please consult subsequent sections). In terms of processing, PU (nano)composite foams are basically prepared according to the methods already presented for both flexible and rigid foams, with the difference that the filler is gener- ally added to one of the basic formulation components, typically the polyol, properly dispersed, in many cases using high-mixing techniques, and at the end. Though some processing particularities have to be considered, such as changes in viscosity and possible modifications in component reactivity, it can be said that PU (nano)composite foams are in general prepared using common PU foaming technologies.

## 6.4    HYBRID POLYURETHANE NANOCOMPOSITE FOAMS

Though polyurethanes present a wide range of physical and chemical properties derived from the specific control of their formulation, enabling their use in almost all types of applications, from coatings or adhesives to composites or even foams, they still present some limitations, such as relatively low thermal stability and mechanical strength. This fact is more marked in the case of PU foams, as foaming, while it improves some of PU's characteristics such as thermal insulation, further decreases its mechanical strength. The incorporation of nanosized reinforcements to PU and later foaming could overcome some of these limitations, as it has been shown that their addition to PU could counteract the loss in mechanical properties inherent to foaming by their mechanical reinforcement and direct effect on the development of a finer and more uniform cellular structure, the combination of which could promote the formation of stronger PU foams, further expanding their possible range of applications. The incorporation of nanosized particles into a polymer matrix and later foaming results in what is known as *polymer nanocomposite foam*. Polymer nanocomposite foams have been receiving increasing attention in the last years, as the addition of nanosized functional particles has been seen as a viable strategy to surpass some of the limitations of these materials while keeping their advantages, such as lightness and thermal insulation (Lee et al., 2005). Recently, this strategy has even been seen as a way to extend the possibilities of these materials by introducing functional nanoparticles that can add specific uncommon characteristics to the foamed material, such as improved flame retardancy or electrical conductivity. Interestingly, foaming has been used in some cases as a strategy to improve the typically poor dispersion of nanometric fillers or particles in polymeric matrices, hence overcoming the main drawback to a wider industrial implementation of these materials and opening up brand new possibilities in the development of lightweight parts for specific applications (Antunes, Velasco, and Realinho, 2011).

Among polymeric foams, PU foams account for the largest market, a direct result of their already mentioned high versatility in terms of final mechanical behavior (from elastomeric to rigid-like), wide density range, and final developed cellular structure and microstructure. Owing to their good thermal insulation, energy absorption, and generally speaking good specific mechanical properties, PU foams are used in the most varied applications, from cushioning and packaging to construction and building, where they are used as cores in sandwich-like structures or even in combination with other structural materials, such as fiber-reinforced composites. In these cases, mechanical strength is a critical asset, especially in the case of PU foams, where the cellular structure is formed by a relatively fast chemical reaction between liquids occurring simultaneously to PU's polymerization (please consult Section 6.2 and Figure 6.1). Though essential in guaranteeing quick foam stabilization, polymerization has to be carefully controlled in order to allow proper foam growth.

For all of the previously mentioned reasons, PU nanocomposite foams could comprise the advantages of PU foams (wide range of densities, versatile mechanical response, reduced thermal conductivity, etc.) with those of incorporating nanosized particles into a polymeric matrix (improved thermal stability and mechanical strength). The nanosized particles most commonly used in PU foams are clearly silicate-layered nanoclays, and particularly unmodified or organically modified montmorillonite (MMT), though others have also been considered, such as carbon-based nanofillers (carbon nanotubes and nanofibers, and more recently graphene), nanosilica, or cellulose-based nanofillers.

This section considers the most relevant works published in the last years about rigid and flexible PU nanocomposite foams, focusing on the influence of the incorporation of different types of nanoparticles on the structural and mechanical, transport, and other relevant properties of the resulting foams.

### 6.4.1 STRUCTURE AND MECHANICAL PROPERTIES

As expected, most of the work published in terms of the analysis of the mechanical properties of PU nanocomposite foams has considered the use of rigid-like PU foams, as these are the most extended in structural applications. Clearly the most considered nanosized reinforcement has been silicate-layered nanoclays, and particularly unmodified or previously modified montmorillonites. For instance, the research group of Widya and Macosko (Cao et al., 2005; Widya and Macosko, 2005) has found that although the addition of 1 wt% organically modified MMT to PU and later foaming resulted in rigid foams with finer cellular structures, no significant improvements were observed in terms of final mechanical strength, which seemed to be supported by other authors using different types of nanometric particles (Javni et al., 2002; Krishnamurthi, Bharadwaj-Somaskandan, and Shutov, 2001).

However, most of the results found in the literature indicate the positive effects that nanoclay incorporation has in terms of improving the mechanical strength of rigid PU foams. Kim et al. (2007) have shown that the incorporation of a fixed concentration of an organically modified nanoclay into a MDI-based PU formulation and later foaming using three different blowing agents (water, cyclopentane, and HFC-365mfc) resulted in rigid PU nanocomposite foams with higher tensile strengths than the unfilled rigid PU foam counterparts. Even higher tensile strength increments were attained by applying ultrasounds to the liquid PU-clay mix prior to its foaming, related to an improved dispersion of the clay layers.

Xu, Tang et al. (2007) have demonstrated that the addition of 2 php of an organically modified nanoclay significantly improved the tensile and compressive strengths of rigid PU foams prepared using water as a blowing agent (around 112% increment in terms of the tensile strength and about 155% in terms of the compressive strength measured at 10% deformation), even at comparatively lower relative densities, related to a combination of the finer cellular structure of PU nanocomposite foams promoted by an effective cell nucleation effect of the well-dispersed clay layers and a higher hydrogen bonding between the PU and clay's organomodifier. Examples of increasingly higher foam cell size reductions by introducing increasingly higher amounts of organically modified MMT (oMMT) can be seen in Figure 6.9. Organoclay was shown to act on its own as a blowing agent during foaming, resulting in PU foams with increasingly lower relative densities while increasing the amount of added oMMT (compare Figure 6.9a,b,c).

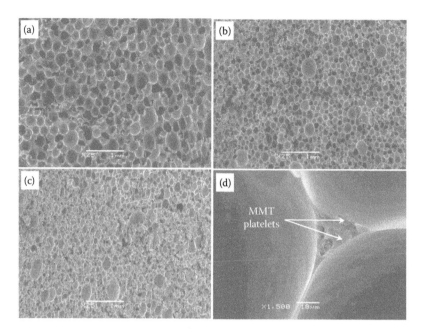

**FIGURE 6.9**  Scanning electron micrographs of: (a) 0.21 relative density unfilled PU foam. (b) 0.23 (1 wt% oMMT). (c) 0.13 (15 wt% oMMT) relative density PU-oMMT foams. (d) Detail of a PU-oMMT foam cell strut showing the presence of MMT platelets.

Mondal and Khakhar (2007) prepared medium-density rigid PU foams with improved compressive and storage moduli by incorporating variable concentrations of different types of unmodified and organically modified montmorillonites, showing promising applications as structural elements. The improvement in mechanical properties was related to the mechanical reinforcement effect of the well-dispersed clay layers and especially the finer foam cellular structure resulting from effective clay layer cell nucleation during foaming. In a similar way, Chen et al. (2011) have recently prepared rigid PU foams with improved mechanical properties through the incorporation of both intercalated and previously calcined modified kaolin, another type of layered silicate. Once again, the better mechanical properties of rigid PU foams with the calcined kaolin were related to their better dispersion in PU when compared to the intercalated kaolin.

Related to the importance of proper clay dispersion in attaining an effective mechanical reinforcement of rigid PU foams, Valizadeh, Rezaei, and Eyvazzadeh (2011) encountered an optimum nanoclay concentration of 1 wt% for compressive strength and moduli improvement of rigid PU foams prepared using a blowing agent mixture of cyclo and normal pentanes. Further nanoclay addition resulted in foams with poorer mechanical properties, which the authors related to a tendency of clay layers to reaggregate at higher contents.

As dispersion and layer intercalation/partial exfoliation have been shown to be critical in terms of attaining the highest possible mechanical reinforcement possibilities of silicate-layered nanoclays, several works have focused on developing different strategies to promote clay layer intercalation/exfoliation and improve surface inter-action with PU. The most used one has been the surface modification of nanoclays prior to mixing with PU's formulation or the combination of previous clay modifica-tion with the application of ultrasounds during mixing. For instance, Seo et al. (2006) have shown that the application of ultrasounds in combination with MDI surface-modified nanoclay resulted in an improved rupture of clay aggregates and intercala-tion of the layers in the PU matrix, improving the efficiency of clay modification and ultimately resulting in PU foams with finer cellular structures and improved flex-ural and tensile strengths. Xu et al. (2010) have specifically studied how the surface modification of MMT using different organomodifiers (cetyltrimethyl ammonium bromide, methyl tallow bis(2-hydroxyethyl) quaternary ammonium chloride and tris(hydroxymethyl)aminomethane) and later incorporation into a PU foam formula-tion affected the final properties of the resulting rigid PU foams. Modification using organomodifiers that contained hydroxyl groups that could react with isocyanate led to partial exfoliation of the clay layers in the PU foams, ultimately resulting in foams with finer cellular structures and higher specific compressive strengths at relatively high fillers amounts, once again demonstrating the importance of dispersion and clay layer exfoliation. Sarier and Onder (2010) have analyzed the possible use of MMT particles modified with different types of low molecular weight polyethylene glycols as mechanical reinforcements in rigid PU foams. Comparatively, PU foams with 2 wt% of modified MMT particles presented a higher storage elastic modulus than the unfilled PU foams, which was related to a good dispersion of the MMT layers throughout the PU and improved interaction of the modified layers with PU's molecules.

Other strategies for improving clay exfoliation in rigid PU foams include the addition of coupling agents, mainly silane-based (Han et al., 2008; Kim et al., 2008), or processing modifications, in addition to the already mentioned ultrasounds, such as microwave processing (Lorenzetti et al., 2010).

A great deal of interest has been given recently to the development of PU nano-composite foams synthesized from isocyanates/polyols obtained from natural resources, mainly derived from the need to replace the increasingly more expen-sive polyols and isocyanates obtained from petroleum. Biobased polyols obtained from oils extracted from plants, such as soybean oil, are currently the ones with the most promising potential to partially or fully replace conventional petrochemi-cal polyols in PU foam formulations, as has been reported in recent years (Banik and Sain, 2008; Guo et al., 2002). However, the use of soybean oil-based polyols, which typically present a considerably lower number of reactive hydroxyl groups than petrochemical polyols (Narine et al., 2007), tends to result in rigid PU foams with lower loading properties and compressive strengths when compared to conven-tional petroleum-based foams. The incorporation of high surface area nanoparticles could counteract these limitations, enabling obtaining considerably cheaper rigid PU foams for the most varied applications.

Liu, Petrovic, and Xu (2009) prepared clay-reinforced PU foams obtained using a soybean oil-based polyol, focusing on the analysis of the effects of incorporating different types and concentrations of organically modified nanoclays. They showed that a clay content with 2.5 wt% the compressive mechanical properties of the foams were kept constant while decreasing density, which was mainly related to a cell density increase and cell size decrease in the foams promoted by clay addition, which could ultimately have beneficial results in improving the thermal insulation properties of the PU nanocomposite foams.

Liang and Shi (2011) prepared rigid PU nanocomposite foams from MDI and soybean-based polyols by incorporating from 0.5 to 3 php of nanoclay. Both the compressive strength and modulus increased significantly for a 0.5 php concentration of clay (respectively 98 and 26% increments when compared to the unfilled soybean-oil based PU foam), progressively decreasing by further increasing the amount of nanoclay. Both tendencies were related in the case of the first to a proper clay dispersion and high cell density and small cell size of the resulting PU foams, and in the case of the higher clay amounts to insufficient nanoparticle dispersion and as a consequence nonuniform foam cellular structure.

Recently, Fan et al. (2012) prepared and characterized rigid PU composite foams containing 15% of soybean-based polyol reinforced with variable concentrations (from 1 to 7 wt%) of hollow glass microspheres and unmodified montmorillonite nanoclays. Though from 1 to 3 wt% glass microsphere content, the compressive strength of PU foams decreased slightly, it increased up until 7 wt%, reaching at this microsphere content a compressive strength comparable to that of unfilled PU foams made from 100% petroleum-based polyol. On the contrary, PU composite foams reinforced with 7 wt% nanoclay presented lower compressive strengths, related to the lower density of these foams and partially opened cell walls promoted by the clay nanoparticles. However, an analysis of the specific compressive modulus showed that PU foams reinforced with 5 to 7 wt% microspheres and 3 to 7 wt% nanoclay presented similar compressive properties than those of unfilled PU foams, mainly explained by the finer cellular structure of the filled foams.

Other natural-based oils have also been considered in the preparation of PU nanocomposite foams, such as palm oil. For instance, Chuayjuljit, Maungchareon, and Saravari (2010) considered the preparation of palm oil-based PU foams reinforced with MMT by *in situ* polymerization. It was found that the compressive strength of PU foams with densities around 40 kg/m³ significantly increased with increasing the concentration of MMT, which was related to an effective mechanical reinforcement effect of the foam promoted by the properly dispersed clay layers and the finer cellular structure of the produced foams.

Javni et al. (2011) studied the possibility of replacing polyol copolymers used in the preparation of flexible PU foams by the incorporation of unmodified and modified montmorillonites. While the addition of the unmodified MMT increased hardness, compression strength, and resilience of the foams, the incorporation of the organically modified MMT resulted in foams with lower modulus, hardness, and compression strength values, a direct result of the higher open-cell contents and poorer cellular structure of the resulting foams, apparently leading to the conclusion that in this specific case the addition of modified MMT led to a counterproductive result.

As a matter of fact, in some specific cases nanoclays have been added to PU foams with the objective of acting as cell open promoters during foaming. As an example, Harikrishnan, Patro, and Khakhar (2006) have used unmodified and organomodified

MMT clays to promote the formation of both rigid and flexible open-cell PU foams. They observed that both types of clays, especially the organically modified one, efficiently acted as cell openers during foam growth, with open cell content increasing with increasing the amount of added MMT. A direct consequence of their higher open cell contents, flexible PU-MMT foams presented improved softness, while rigid PU-MMT foams displayed a higher dimensional stability.

Though silicate-layered nanoclays have clearly been the most used nanosized reinforcement in the preparation of PU nanocomposite foams, a great number of researchers have recently considered the addition of carbon-based nanoparticles, particularly carbon nanotubes (CNTs), carbon nanofibers (CNFs), and graphene, mainly driven by the intrinsically high mechanical and especially high transport properties of these materials, which have opened up a new set of possibilities in sectors such as electronics (Shaffer and Sandler 2007; Singh et al., 2011).

Dolomanova et al. (2011) recently studied the mechanical properties of *in situ* polymerized rigid PU foams reinforced with 0.5 and 1 wt% of unmodified and diphenylmethane 4,4'-diisocyanate-modified carbon nanofibers (CNFs) and single and multi-walled carbon nanotubes (SWNTs and MWNTs, respectively). They observed that the incorporation of the modified MWNTs resulted in a mechanical improvement around 94 and 70%, respectively, in terms of the specific elastic modulus and specific compressive strength, with regard to the unfilled PU foam. The mechanical reinforcement effect of adding MWNTs was higher than that of SWNTs due to the lower tendency of MWNTs to reaggregate during processing and higher than the CNFs due to their higher specific surface interaction area. MWNT surface modification enhanced its interaction with the polymer matrix, promoting a better dispersion of the filler, which led to the formation of PU foams with finer cellular structures and higher mechanical properties. Further investigation by these researchers showed the importance of processing on the mechanical properties of the resulting foams (Zhang et al., 2011). In addition to surface modifying the MWNTs prior to their incorporation into the polyol, high-shear mixing was used in order to obtain a proper dispersion of the nanotubes. Interestingly, the reinforcement effect of the MWNTs was more significant with increasing high-shear mixing time than with increasing the concentration of nanotubes, with no significant mechanical reinforcement effect being observed by functionalizing the carbon nanotubes.

Yan et al. (2011) prepared rigid PU nanocomposite foams reinforced with variable concentrations of carbon nanotubes for long-term use electrical conductive components. Particularly, for a 2 wt% CNT content, rigid PU foams presented around a 30% increase in compression properties and a 50% increase in storage modulus, both measured at room temperature, when compared to the unfilled PU foam, thus demonstrating the effective mechanical reinforcement effect that low amounts of carbon nanotubes have on PU foams.

In a similar way as when using nanoclays (Liang and Shi, 2011), Liang and Shi (2010) considered the incorporation of 0.5 and 1 wt% of MWNTs to improve the mechanical performance of PU foams obtained from soybean-based polyols. They observed that MWNT addition resulted in rigid PU foams that presented improvements higher than 20% in terms of the compressive, flexural, and tensile mechanical properties when compared to the unfilled PU foam.

Saha, Kabir, and Jeelani (2008) considered the incorporation of different types of nanoparticles at 1 wt% content into a rigid PU foam formulation and mixing using ultrasonication. Among these, carbon nanofibers (CNFs) had the most remarkable effect on the mechanical properties of PU foams, improving the elastic modulus about 86, 40, and 45%, respectively, in tension, compression, and flexion, and mechanical strength about 35, 57, and 40% with regard to the unfilled reference PU foam. The same authors had demonstrated in a previous publication the importance of applying an adequate ultrasonication time when mixing the carbon nanofibers with the polyol component (Kabir, Saha, and Jeelani, 2007). Under optimized ultrasonication mixing conditions the incorporation of 0.5 wt% CNF resulted in an improvement of the compressive strength that was slightly higher than 10% when compared to the unfilled PU foam, further increasing to 21% for a CNF content of 1 wt%. Despite the applied ultrasonication treatment, the compressive strength of PU foams decreased when adding 1.5 wt% CNF, related to a certain reaggregation of the carbon nanofibers during processing.

Though very limited work has been dedicated to the incorporation of graphene or graphite nanoplatelets into PU foams, some researchers have recently considered these carbon-based layered nanoparticles as promising materials for attaining foam properties that can be seen as those of combining some of the advantages of nanoclays (improved specific surface area related to their platelet-like structure, which could result in some added functionalities such as improved barrier properties) and carbon nanotubes/nanofibers (high transport properties). For instance, Chen, Yang, and Chen (2010) have studied the addition of graphene nanosheets as possible replacers of common carbon black as conductive filler in PU foams. Not only did they show that PU foams with graphene nanosheets presented a lower electrical percolation threshold than those with carbon black, but they also presented a higher tensile strength (around 3.5 MPa for PU-graphene foams compared to a bit less than 1 MPa for the ones with carbon black for a 3 wt% filler content) related to the higher aspect ratio of the graphene nanoplatelets and improved dispersion throughout PU foam's cell walls.

Yan et al. (2012) have compared the mechanical properties of rigid PU foams reinforced with graphene nanoplatelets and MWNTs. The addition of only 0.3 wt% of graphene and MWNTs led to PU foams with, respectively, 36 and 25% higher compressive modulus when compared to the unfilled rigid PU foam. The higher mechanical reinforcement effect of graphene at the same concentration when compared to MWNTs was once again attributed to the particular platelet-like surface of graphene, which presented a higher specific surface area of interaction with PU's matrix than the nanotubes. In both cases, it was found that the incorporation of small amounts of carbon nanofillers could greatly improve the mechanical response of rigid PU foams and thus significantly expand their range of possible applications.

Owing to the particular high transport properties of carbon-based nanofillers, a great number of recent publications have considered the incorporation of these nanofillers into flexible PU foams, as opposed to silicate-layered nanoclays, which are mainly added as mechanical reinforcements and for that reason are almost only considered for rigid PU foams. The addition of carbon-based nanofillers comes from the interest in developing new functional flexible foams, for instance with improved piezoresistivity properties.

The group of Verdejo has analyzed the effect of both unmodified and functionalized MWNTs on the *in situ* foaming evolution of flexible PU foams (Bernal, López-Manchado, and Verdejo, 2011), as well as the effect of unmodified and modified MWNTs and graphene obtained from the reduction of graphite oxide on their physical properties (Bernal et al., 2012). As expected, the addition of MWNTs, especially the functionalized ones, resulted in PU foams with improved storage modulus with increasing the amount of nanotubes, demonstrating the effective mechanical reinforcement effect of these nanofillers. Improved thermal stability as well as enhancement of the electromagnetic interference (EMI) shielding effectiveness was observed for the nanocomposite foams due to the presence of the nanoparticles. Comparatively, the incorporation of a fixed amount of 0.3 wt% of functionalized graphene had a greater effect in improving both the thermal stability and EMI shielding effectiveness than both unmodified and modified MWNTs, which was mainly related to the higher aspect ratio of graphene, greatly promoting the formation of an effective conductive network in the flexible PU foams.

Contrary to the group of Verdejo, which had only considered the use of oxidized MWNTs, Bandarian, Shojaei, and Rashidi (2011) have analyzed the influence of the addition of 0.1 wt% of different types of highly functionalized MWNTs, particularly having carboxyl, hydroxyl, and amide functional groups, on the properties of open-cell flexible PU foams. They observed that the mechanical properties of PU foams were greatly improved with incorporating all of the functionalized MWNTs, with the ones modified with carboxyl groups showing the highest enhancement (approximately 50% improvement in both tensile modulus and strength, and 40 and 60% enhancements, respectively, in compressive modulus and compressive strength, when compared to the unfilled PU foam), which was related to a better interfacial interaction with the polymer matrix and improved dispersion of these nanotubes (consult Table 6.2 for further details).

Several studies have also shown an improvement of the mechanical properties of rigid PU foams by adding small amounts of spherical-like nanosilica (Fan and Xiao-Qing, 2009; Wang et al., 2004; Xie and Wang, 2005). Particularly, Nikje and Tehrani (2011) have incorporated two different types of silane-based surface-modified

**TABLE 6.2**
**Mechanical Properties of the Unfilled and MWNT-Reinforced PU Foams**

| Material | MWNT Functionalization | Tensile Modulus (kPa) | Tensile Strength (kPa) | Compressive Modulus (kPa) | Compressive Strength (kPa)[*] |
|---|---|---|---|---|---|
| Unfilled PU | — | 0.66 | 104 | 0.13 | 3.2 |
| PU/MWNT-NH | Amide | 0.75 | 129 | 0.14 | 3.7 |
| PU/MWNT-OH | Hydroxyl | 0.97 | 154 | 0.18 | 4.6 |
| PU/MWNT-COOH | Carboxyl | 0.98 | 156 | 0.18 | 5.1 |

*Source:* Prepared by Bandarian et al., 2011, *Polym. Int.* 60, 475–482.
[*] Measured at the plateau of the curve.

nanosilicas and checked that both globally improved the mechanical properties of rigid PU foams, greatly related to an improved dispersion of the nanosilica particles in the PU matrix promoted by the silane coupling agents.

Finally, in the line of developing 100% natural-based PU foams with improved mechanical performance, cellulose nanowhiskers have been considered as a possible mechanical reinforcement of PU foams synthesized from polyols obtained from renewable sources, particularly vegetable oils. The research group of Li et al. (2009) have synthesized and characterized rigid PU foams reinforced with cellulose whiskers (Li, Ren, and Ragauskas, 2010, 2011), demonstrating that the addition of 1 wt% of cellulose resulted in increments in terms of the compressive strength and modulus of, respectively, 270 and 210% with regard to the unfilled PU foam. Luo, Mohanty, and Misra (2012) used microcrystalline cellulose as a mechanical reinforcement of PU foams obtained from both soybean-derived and petrochemical-obtained polyols, showing that the compressive and flexural strengths and moduli were significantly improved with increasing the amount of added cellulose. The results presented in this work show the potential of using nanocellulose as a mechanical reinforcement for the development of partially or fully biobased PU foams.

### 6.4.2 Transport Properties

Though the incorporation of nanoparticles into PU foams has mainly been seen as a way to counterbalance the usual reduction in mechanical strength resulting from foaming, hence extending their possible range of applications, especially in the case of rigid PU foams, a great deal of interest has been given to the improvement of the transport properties of PU foams by introducing conductive nanoparticles. In this field, special emphasis has been given to the addition of conductive carbon-based nanofillers, and among them, carbon nanotubes, carbon nanofibers, and platelet-like graphite or graphene.

Owing to the extremely high electrical conductivity of carbon-based nanofillers, their incorporation in low amounts to both rigid and flexible PU foams has been seen as a strategy to generate electrically conductive lightweight components for the most varied applications, from static dissipation to EMI shielding (Antunes and Velasco, under review). Recently, Athanasopoulos et al. (2012) have analyzed the electrical conductivity of medium-density rigid PU nanocomposite foams with different amounts of unmodified, that is, as-received MWNTs. They demonstrated that it was possible to prepare electrically conductive rigid PU foams over a wide range of foam densities (from 250 kg/m$^3$ to as high as 400 kg/m$^3$) and nanofiller contents (1, 2, 3, and 5 wt% MWNT), with the foams presenting a characteristic electrically conductive percolative behavior. It was found that higher density PU foams already presented a significant increase in terms of electrical conductivity at 1 wt% MWNT, with the electrical conductivity increasing from 10$^{-12}$ S/m of unfilled PU (insulator behavior) to 10$^{-8}$ S/m (slightly conductive behavior). Increase of MWNT's content to 2 wt% further increased the electrical conductivity by four orders of magnitude (10$^{-4}$ S/m), with much milder enhancements being observed with further increasing of the amount of MWNTs up to 5 wt%, reaching an electrical conductivity plateau at around 10$^{-2}$ S/m. A similar trend was observed for the lower density PU foams, though the conductivity only started increasing at 2 wt% MWNT and the rate of electrical conductivity increase was considerably smaller than that observed for the denser foams, indicating that the critical MWNT concentration for electrical conduction greatly depends on density. In any case, both denser and lighter PU-MWNT nanocomposite foams presented an electrical behavior that followed a typical percolation model, with foams presenting an insulator to conductor transition at specific nanofiller content, related to the formation of an electrically conductive network by direct contact between the nanotubes.

Xu, Li et al. had already shown in 2007 the importance of density in the electrical conductivity of PU foams reinforced with 2 wt% of carbon nanotubes, observing an insulator to conductor transition behavior that resembled that of percolative polymer-CNT composites (Xu, Li et al., 2007). Similarly, Xiang et al. (2009) analyzed the electrical behavior of medium-density rigid PU foams with a fixed concentration of MWNTs, focusing on the temperature dependence of the electrical resistivity of the foams. These authors observed that resistivity decreased with heating and reheating the foams between 25 and 100°C, related to an increase in the pressure of the $CO_2$ enclosed inside the cells. As more recently observed by Athanasopoulos et al. (2012), the increase in foam density for this MWNT concentration favored a closer distance between adjacent nanotubes, resulting in more effective conductive paths and hence higher electrical conductivities.

You et al. (2011) considered the preparation and later characterization of the electrical conductivity of MWNT-reinforced PU foams. They found that the electrical conductivity of the resulting foams followed a percolative behavior with increasing the amount of carbon nanotubes until an MWNT content of 0.1 php, with the electrical conductivity increasing dramatically from the almost 0 of the unfilled PU foams (typical insulating behavior) to the 0.23 S/cm of the PU nanocomposite foams with 0.1 php MWNT (semiconducting behavior). Further MWNT addition until a maximum content of 0.5 php did not significantly alter the electrical conductivity, which was related to MWNT aggregation at concentrations > 0.1 php and thus impoverished electrical conduction efficiency.

In addition to analyzing the effect of different amounts of MWNTs (from 1 to 2 wt%) on the compression properties of rigid PU foams, Yan et al. (2011) also studied the electrical conductivity of the resulting foams for a fixed density of 200 kg/m³. They observed that PU-MWNT foams followed a typical electrically conductive percolative behavior, in this case with an electrical percolation threshold of 1.2 wt% MWNT, once again related to the formation of an effective MWNT conductive network by homogeneous dispersion of the nanotubes throughout the cell walls and struts of the PU foam (see the scheme presented in Figure 6.10). Particularly, the typical insulating behavior of PU foams changed to a semiconducting one with the addition of 1.2 wt% MWNTs, with the electrical conductivity increasing in more than six orders of magnitude until reaching a value of $2.03 \times 10^{-6}$ S/m. Also, the conductive PU-MWNT foams presented great electrical stabilities over a wide range of temperatures (from 20 to as high as 180°C), thus enabling their use in long-term use electrically conductive applications.

As previously mentioned, recent research in the field of PU nanocomposite foams has concentrated in the development of novel piezoresistive flexible foams

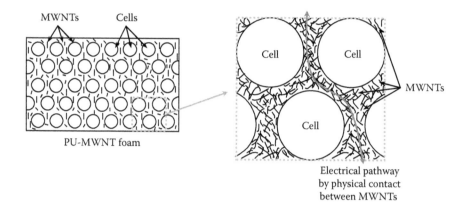

**FIGURE 6.10** Scheme of a PU-MWNT nanocomposite foam showing the formation of an effective electrical pathway by physical contact between MWNTs homogeneously dispersed throughout the foam cell walls and struts.

by combining flexible PU foam formulations with conductive carbon nanofillers. Some patents have even considered the preparation of soft and flexible PU foams with improved electrical conductivity properties by adding conductive carbon-based nanoparticles, especially carbon nanotubes and nanofibers (Li and Xiang, 2009).

Dai et al. (2012) have studied the variation in electrical conductivity of PU foams reinforced with MWNTs upon compression. These authors observed that volume resistivity was mainly a function of the level of damage inflicted to foam struts during compression. Once a stable electrically conductive microstructure was formed in the foams by direct contact of the nanotubes, these presented a constant value of volume resistivity independently of the applied external mechanical stimulus. Hodlur and Rabinal (2012) have recently presented a work that addressed the preparation of conductive flexible PU foams by initially introducing graphite oxide and later chemically reducing it to conductive graphene. They studied the pressure dependent electrical conductivity of these foams and showed that the electrical conductivity could be increased by five orders of magnitude by applying pressure, which was related to the formation of an effective electrical conduction path by direct contact between the graphene platelets.

These recent results show promising possibilities in the development of new piezoresistive sensors based on flexible PU nanocomposite foams with low amounts of conductive carbon-based nanoparticles.

As one of the main advantages of foaming a PU-based matrix is the reduction in thermal conductivity, thus explaining the vast use of these materials as thermal insulators. A great number of publications have considered the effect that the addition of nanoparticles may have on the thermal conductivity of both rigid and flexible PU foams. In their work, Kim et al. (2007) studied the effects of incorporating a fixed concentration of organically modified nanoclay and ultrasound mixing on the thermal insulating behavior of rigid PU foams prepared using CFC-free blowing agents. As the application of ultrasounds resulted in PU-MMT foams with considerably smaller cell sizes, a direct result of an improved dispersion of the clay layers and hence improved heterogeneous cell nucleation effect, PU nanocomposite foams obtained using ultrasounds presented the lowest thermal conductivity. Therefore, clay addition may be seen under specific conditions as a possible strategy to further decrease the thermal conductivity of PU foams, thus enabling the reduction of the required material thickness for a given thermal insulating application.

Interestingly, the incorporation of low amounts of conductive carbon nanoparticles to PU and later foaming has been shown by several authors to result in electrically conductive lightweight materials with thermal insulating properties, explained by the different mechanisms involved in electrical and thermal conductions. For instance, You et al. (2011) observed that the addition of low amounts of MWNTs resulted in thermal insulating PU foams, with these foams presenting even lower thermal conductivities than those of foams with higher MWNT contents, related to their finer cellular structure resulting from a much more effective heterogeneous cell nucleation effect promoted by the well-dispersed carbon nanotubes.

### 6.4.3 OTHER PROPERTIES

In addition to mechanical and transport properties, other relevant properties of both flexible and rigid PU foams have also been addressed. Among these, flame retardancy, derived from the increasingly more stringent fire retardancy standards that have emerged in sectors such as construction or automotive, where PU foams, especially of the rigid type, have a great importance, as well as sound damping in the case of flexible foams, have been greatly considered.

An extensive review of the fire retardancy of both flexible and rigid PU foams has been presented by Singh and Jain (2009). As most polymer foams, one of the main drawbacks of PU foams derives from their high flammability and, in the specific case of PU, their high amount and toxicity of dark smoke generated during combustion, considerably limiting their use in many applications. Though the addition of different compounds, mainly phosphorous-containing, halogen-containing, nitrogen-containing, or silicone-containing products have been proven to render fire retardancy to PU foams and thus have been widely used throughout the years, recent fire retardancy standards are considerably limiting or even prohibiting their use. In addition, their incorporation tends to considerably reduce the mechanical properties as well as directly affect other important physical characteristics of the resulting foams. For these reasons, a great deal of attention has been given to the possibility of improving the inherently low fire retardancy of PU foams by incorporating low nanoparticle content, alone or in combination with other flame retardant systems. Gilman's group has recently considered the addition of both nanosized clays and carbon nanofibers as a strategy to improve the fire resistance of flexible PU foams, mainly focusing on the heat release rate (HRR) generated during burning and melt dripping (Zammarano et al., 2008). Interestingly, these authors found that carbon nanofibers formed an entangled network at the surface of the foam during combustion in a cone calorimeter, eliminating melt dripping and globally reducing the HRR. In a similar way, Harikrishnan et al. (2010) showed that the addition of only 1 wt% carbon nanofibers significantly improved the fire retardancy of PU foams by increasing the level of weight retention after the removal of the flame in 13%.

Nanosized particles have also been considered in combination with other flame retardant systems as a strategy to reduce the flammability of PU foams. For instance, Modesti et al. (2008) used unmodified and organically modified silicate layered nanoclays in combination with aluminium phosphinate. Phosphinate was shown on its own to significantly reduce the flammability of PU foams by acting in both condensed and gas phases during combustion. While the addition of both pristine and ammonium-modified nanoclays did not have a significant effect in decreasing the peak of heat release rate (PHRR) and total heat release (THR) during cone calorimeter burning, only acting as a surface physical barrier to the diffusion of combustion volatiles and oxygen, the incorporation of diphosphonium-modified nanoclay in combination with aluminium phosphinate resulted in an important reduction of the PHRR and THR, which was related to the synergistic fire retardant effect of the phosphonium clay modifier, that also acted in both condensed and gas phases.

As previously mentioned, sound damping is one of the properties more commonly required for flexible PU foams. Sung et al. (2007) have successfully improved the sound damping properties of open-cell flexible PU foams by adding unmodified and organically modified platelet-like montmorillonites. Similarly, Verdejo et al. (2009) found that the addition of up to 0.1 wt% carbon nanotubes resulted in open-cell flexible PU foams with improved sound absorption over the whole analyzed frequency range (1000–2000 Hz), increasing the absorption from 70% of unfilled PU foam to as high as 90%. This spectacular increase in sound absorption was explained by the authors by the large surface area at the PU–CNT interface, where energy could be more easily dissipated.

Bandarian, Shojaei, and Rashidi (2011) found out that the incorporation of reactive modified MWNTs, particularly the ones with hydroxyl and carboxyl functional groups, helped in improving the sound absorption properties of flexible PU foams by inducing the formation of microcells of less than 5 μm in the open-cell walls of the foams.

Very recently, Willemsen (2012) has shown that the inclusion of different types of MWNTs can significantly improve the sound damping properties of flexible PU foams, with results depending on both MWNT particle size and concentration, thus showing the great possibilities of nanoparticle incorporation as a strategy to improve the sound absorption properties of open-cell flexible PU foams.

## 6.5 NEW DEVELOPMENTS IN HYBRID POLYURETHANE NANOCOMPOSITE FOAMS

The incorporation into PU foams of different types and concentrations of nanosized functional particles, both platelet-like such as nanoclays or graphene, fibrous-like such as carbon nanotubes and nanofibers, or even spherical-like such as nanosilica, has been shown to have a significant effect in the structure and physical properties of the resulting materials, considerably expanding the already wide range of possible applications of these lightweight materials.

In addition to the already mentioned effects in terms of attaining both flexible and rigid PU foams with improved mechanical performance, enhanced electrical conductivity, or even improved fire retardancy or sound damping, depending on the type of added nanoparticle and developed foam cellular structure, new developments in hybrid PU nanocomposite foams have emerged in recent years, mainly a consequence of the development of novel nanosized materials, better understanding of the characteristics and properties of these materials, as well as improvement of the methods used to adequately incorporate them into polymer-based systems.

Taking advantage of the particular platelet-like geometry of nanoclays and their high ion-exchange capability, intercalation, and swelling properties, different studies have appeared very recently related to the preparation of nanohybrids made from carbon nanotubes synthesized in the surface of montmorillonite nanoclays (MMT-CNT nanohybrids) (Gournis et al., 2002; Li et al., 2009; Lu et al., 2005; Madaleno et al., 2012b; Zhang, Phang, and Liu, 2006). Owing to the particular combination of properties of both nanofillers, the construction of these nanohybrids and later incorporation into PU foams could result in remarkable improvements in mechanical and physical properties.

Pyrz's research group has just published a couple of works about the preparation of MMT-CNT nanohybrids and their incorporation into PU foams (Madaleno et al., 2012a, 2013). The nanohybrids were prepared by chemical vapor deposition (CVD) of the carbon nanotubes on the surface of different types of iron-montmorillonite nanoclays. It was observed that the nanotubes formed a network structure connecting the clay layers and helped to increase clay interlayer space (see the scheme presented in Figure 6.11). The synthesized nanohybrids were incorporated into rigid PU foam formulations and PU nanocomposite foams were prepared by *in situ* polymerization. PU nanocomposite foams presented considerably finer cellular structures when compared to the unfilled PU foam, which was explained on the basis of an effective cell nucleation effect promoted by the properly dispersed nanohybrids, which in turn was favored by the pre-exfoliated internal structure of MMT assisted by the surface growth of carbon nanotubes. Not only the addition of the nanohybrids improved the thermal stability of the resulting foams, but most importantly they significantly enhanced the compressive mechanical properties of PU nanocomposite foams, offering new possibilities for their use in several structural applications. Particularly, significant improvements in terms of the specific compressive modulus and strength were attained with only adding a 0.25 wt% of nanohybrid (respectively

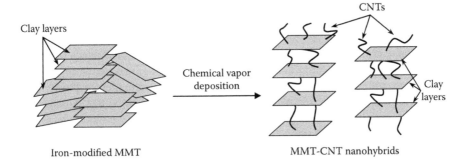

Iron-modified MMT                                MMT-CNT nanohybrids

**FIGURE 6.11**  Basic scheme showing the formation of MMT-CNT nanohybrids according to Madaleno L. et al. (From Madaleno L. et al., 2012a, *Polym. & Polym. Comp.* 20, 693–700.)

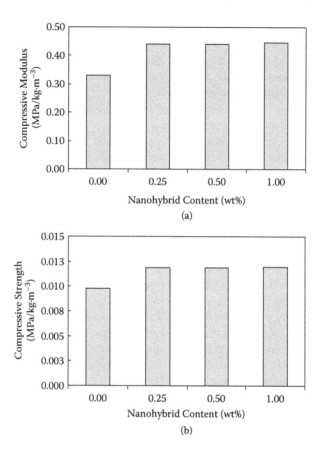

**FIGURE 6.12** (a) Specific compressive modulus and (b) specific compressive strength of unfilled PU foams and PU nanocomposite foams reinforced with MMT-CNT nanohybrids. (Adapted from Madaleno, L. et al., 2013, *Comp: Part A* 44, 1–7.)

31 and 20% improvements when compared to the unfilled PU foams). No further enhancements were observed with increasing the concentration of nanohybrid (see results presented in Figure 6.12) related to the presence of nanohybrid aggregates at higher concentrations.

Based on the inherent self-repair capability of the human body, tissue engineering has emerged from medical sciences in the last 20–25 years as almost a scientific field on its own. Tissue engineering may be generically defined as the *in vitro* or *ex vivo* regeneration of living tissue resulting from the combination of biocompatible structural scaffolds and implanted living tissue cells (Velasco and Antunes, in press). It has been shown that a good design of the cell-supporting structural scaffold is as important as the implanted cells, hence the great amount of effort dedicated to the development of adequate structural scaffolds for tissue engineering applications. These scaffolds have to be biocompatible and biodegradable, highly porous and interconnected in order to allow an effective cell adhesion and penetration, as well as mechanically consistent. Though different materials have been considered in the design of tissue engineering scaffolds, owing to their lightness and mechanical properties, polymer-based materials are clearly the most popular ones. The most used polymers for tissue engineering scaffolds are based on biocompatible synthetic polyesters such as poly(lactic acid) (PLA), poly(glycolic acid) (PGA), polycaprolactone (PCL), and their blends or copolymers (Seal, Otero, and Panitch, 2001). However, segmented PU elastomers, which have been shown to display an adequate combination of biocompatibility and excellent mechanical properties (Christenson, Anderson, and Hittner, 2007; Grenier et al., 2009; Piticescu, Popescu, and Buruiana, 2012), have also been considered in a great number of biomedical applications. PU represents an interesting material for developing tissue engineering scaffolds, as it displays a great versatility in terms of final foam characteristics, enabling its use from soft tissue regeneration (flexible foams) to cartilage and bone repair (rigid foams) (Fromstein and Woodhouse, 2002; Zhang et al., 2003). In this last case, the bioactivity of PU scaffold's surface must be improved in order to properly bond with bone tissue and promote new bone formation, which is usually done by functionalizing its surface prior to bioactive molecule immobilization (Huang, Wang, and Luo, 2009; Jozwiak, Kielty, and Black, 2008).

In the line of the development of new nanohybrids based on the combination of MMT and CNT and incorporation into PU foams for mechanical improvement, CNTs have increasingly been considered in recent years to enhance some properties and functionalities of structural scaffolds for tissue engineering (Harrison and Attala, 2007; MacDonald et al., 2005), such as osteoconductivity and mineralization potential, mechanical reinforcement, or even electrical conductivity, which can be used to promote cell growth. Thus, the development of last generation hybrid scaffolds based on PU foams functionalized with conductive carbon-based nanofillers combined with bioactive micro- or nanometric-sized ceramic particles and/or natural polymers (such as collagen) has generated a great deal of interest (see scheme presented in Figure 6.13).

Zawadzak et al. have considered the development of new tissue engineering scaffolds based on highly porous and interconnected PU foams electrophoretically coated with carbon nanotubes (Zawadzak et al., 2009). These authors observed that the presence of a uniform CNT coating on the surface of PU foams accelerated the formation of calcium phosphate, an indication of the material effectiveness to form strong bonds to bone tissue, when compared to the uncoated foams, which was related to favored calcium phosphate crystal nucleation and growth induced

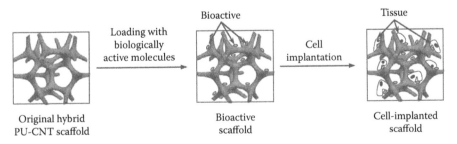

**FIGURE 6.13**  Basic scheme of the use of a hybrid PU-CNT scaffold for cell implantation.

by the carbon nanotubes. In conclusion, hybrid scaffolds based on CNT-coated PU foams showed a great potential to be used in bone tissue engineering, owing to their combination of high porosity and interconnected porous structure, bioactivity, and functionalized nanostructured surface topography.

Similarly, Jell et al. (2008) have prepared hybrid PU-based scaffolds with MWNTs, though they incorporated the carbon nanotubes in the bulk by thermally induced phase separation. Though a good MWNT dispersion throughout the PU matrix could be achieved, a rough nanostructured texture was formed at the surface of the scaffolds, which in fact could further enhance cellular adhesion. As a result, the addition of increasingly higher concentrations of nanotubes altered the surface chemistry of the scaffolds without affecting their wettability while significantly increasing the compressive strength (from 55 $kPa/g{\cdot}cm^{-3}$ of unfilled PU foams to as high as 170 $kPa/g{\cdot}cm^{-3}$ for a MWNT content of 5 wt%). These authors also analyzed the *in vitro* cellular response of the hybrid scaffolds in terms of cytotoxicity, mineralization, and angiogenic potential, observing that the incorporation of increasingly higher amounts of MWNTs did not result in osteoblast-like cells (SaOS-2) cytotoxicity or decreased mineralization. Interestingly, they observed that it was possible to control bone cellular growth by varying the concentration of MWNTs, as assessed by the improvement of the angiogenic factor of vascular endothelial growth after 3 days in culture with increasing the proportion of nanotubes.

As the overall performance of hybrid PU-CNT scaffolds is often hindered by the poor intrinsic quality of carbon nanotubes or their later poor dispersion throughout the scaffold walls, the same research group further extended their study by incorporating chemically modified MWNTs and analyzing the interaction of the resulting hybrid scaffolds with osteoblasts (Verdejo et al., 2009). The incorporation of small concentrations of oxidized MWNTs, proven to be well dispersed throughout the solid matrix due to covalent bonding to the isocyanate component, significantly reduced PU's surface hydrophobicity, providing improved wettability and supporting the claim that nanotubes are mainly active at the surface of the scaffold. The resulting nanocomposite foams adequately supported bone cell adhesion, growth, mineralization, and angiogenesis, with the particularity of qualitatively displaying a better osteoblast SaOS-2 mineralization than unfilled PU scaffolds.

In conclusion, hybrid PU-CNT nanocomposite foams are promising materials for tissue engineering scaffold applications, as they combine the versatility of PU foams in terms of general mechanical behavior (flexible or rigid) and developed cellular structure with the advantages of incorporating small amounts of carbon nanotubes, which can enhance mechanical performance or electrical conductivity. In addition, PU-CNT foams have been shown to support bone cell adhesion and growth, with carbon nanotubes having an important effect in adequately adapting the surface characteristics of the scaffold to specific tissue engineering applications as well as improving cell phenotype.

Due to the particular characteristics of PU foaming, which basically consists of the reaction between two liquids at room temperature, thus enabling the relatively easy incorporation of solid mechanical reinforcements, the research group in which I am involved has recently considered the preparation of novel hybrid rigid PU-based nanocomposite foams with improved properties by applying the concept of multiscalar mechanical reinforcement (Antunes et al., 2011; Antunes, Maspoch, and Velasco, 2012). Generally speaking, multiscalar reinforcement may be defined as the combination of at least two different mechanical reinforcements having different scale dimensions. The multiscalar reinforcement effect of combining nanometric-sized functional reinforcements such as nanoclays or carbon nanotubes with more conventional micrometric-sized fibrous-like reinforcements could not only solve most of the problems related to poor mechanical performance of PU foams, but could also add specific functionalities or characteristics to the foam resulting from synergies between both reinforcements.

As there is a great interest in developing environmentally sustainable low cost structural lightweight materials, biobased hybrid PU foams reinforced with macroscopic fibers, fabrics, or wools obtained from renewable sources, mainly cellulose-based, could come as a good low cost possibility. If low amounts of a nanosized functional material were to be added in combination with the cellulose-based mechanical reinforcement, novel hybrid lightweight materials could be obtained. Particularly, we have considered the simultaneous incorporation of high amounts of macroscopic esparto wool (38–65 wt%), a cellulosic-based material characterized by its high cellulose content, and relatively small amounts of montmorillonite (1–15 wt%) (Antunes et al., 2011). The incorporation of increasingly higher amounts of MMT and especially the multiscalar combination of MMT and esparto resulted in PU foams with considerably finer cellular structures, reaching cell sizes as small as 40 μm for relative densities of 0.25 and even below 100 μm for extremely low relative densities of 0.09, compared to the around 220 μm of unfilled PU foams (see Figure 6.14). The addition of different amounts of MMT can thus be seen as a strategy to regulate the cellular structure of the resulting PU foams in terms of cell size and distribution. However, the addition of both reinforcements resulted in foams with important open-cell contents, a direct result of the well-known cell opening effect of MMT and high esparto content. Though worse in terms of the absolute values of compressive collapse strength derived from their higher open-cell contents, hybrid PU foams reinforced with MMT and esparto displayed a less abrupt decrease of the collapse strength with decreasing relative density, as well as the highest energy absorbed until collapse, hence showing promising possibilities as low cost lightweight structural

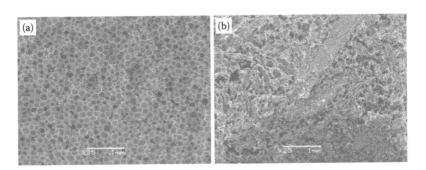

**FIGURE 6.14**  Scanning electron micrographs of (a) unfilled PU and (b) PU-MMT-esparto foams.

materials with tailor-made properties by control of the developed cellular structure (MMT content) and macroscopic mechanical reinforcement (esparto concentration).

Trying to further extend the concept of multiscalar reinforcement applied to rigid PU foams, we have recently presented a work that addressed the combination of organically modified MMT and other cellulose-based reinforcements in addition to esparto, particularly cellulose pulp and cardboard paper (Antunes, Maspoch, and Velasco, 2012). While the incorporation of MMT promoted the formation of PU foams with finer and more homogeneous cellular structures as well as contributed to important density reductions, thus resulting in foams with lower compressive properties than the unfilled ones, the inclusion of both cellulose and cardboard paper significantly increased the collapse strength of PU foams even at lower relative densities, demonstrating their effectiveness as mechanical reinforcements of rigid PU foams. The results presented in both works demonstrate the viability of obtaining rigid PU foams with improved mechanical properties over a wide range of densities by combining the incorporation of low cost macroscopic cellulose-based reinforcements with small amounts of a nanosized filler.

## 6.6 CONCLUSIONS

Polyurethane foams are one of the most successful and most commonly used types of polymer-based cellular materials, owing to their unbeaten combination of good specific mechanical properties and high thermal insulation. Though complex, PU foam formulations have developed in such a way throughout the years that have enabled, along with the emergence of new processing technologies, the preparation of PU foams with a wide range of characteristics for the most varied applications, from open-cell flexible PU foams with improved softness for cushioning to closed-cell rigid foams for structural thermal insulation applications.

Nevertheless, loss in mechanical performance with foaming is still significant, especially in the case of low-density foams, preventing the use of PU foams in a great number of applications. The addition of nanosized particles has recently been considered as a possible strategy to improve the poor mechanical properties of PU foams. As a result of their high specific surface area, nanoparticles have been shown, when properly dispersed, to have a significant effect even at low amounts in mechanically reinforcing PU foams, which has been shown to be mainly a consequence of the combination of the inherently high mechanical properties of said nanoparticles and their influence in generating foam cellular structures with considerably smaller cell sizes and higher cell densities. As a consequence, a great deal of the research work dedicated to PU nanocomposite foams has been focused on low-density rigid foams, as both mechanical reinforcement and improved thermal insulation are to be expected, setting interesting new possibilities for these lightweight materials.

The study and development of novel hybrid PU nanocomposite foams has been strongly guided by the possibility of introducing specific characteristics or properties to PU foams by incorporating functional nanoparticles that combine mechanical reinforcement with other interesting properties. For instance, carbon-based nanoparticles are often added due to their high electrical conductivity, which could result in the emergence of new piezoelectric components based on elastic-like flexible PU nanocomposite foams.

The concept of hybrid PU foams has been taken recently a bit further, mainly a consequence of the arrival of new nanosized materials and nanohybrids, improved methods to incorporate nanoparticles into polymer-based systems, and better understanding of the complex phenomena involved in polymer nanocomposite foaming. Particularly, nanohybrids obtained from the combination of different nanosized particles have been considered as viable materials for the mechanical reinforcement of rigid PU foams at globally lower amounts than common nanoparticles, with the additional advantage of facilitating an improved nanofiller–PU matrix interaction and thus higher nanofiller efficiency. Multiscalar reinforcement by combining low cost macroscopic reinforcements and low amounts of nanoparticles has recently been contemplated as a strategy to improve the mechanical properties of PU foams while decreasing their final cost, thus showing promising possibilities in the building sector. Hybrid PU nanocomposite foams have started to be considered as interesting materials to be used as structural scaffolds for tissue engineering, especially for bone regeneration, though flexible open-cell PU nanocomposite foams are also expected to have possible applications in the regeneration of soft-like tissues.

## ACKNOWLEDGMENTS

The author would like to acknowledge the Spanish Ministry of Economy and Competitiveness for the financial support of project MAT2011-26410.

## REFERENCES

Antunes, M., A. Cano, L. Haurie, and J. I. Velasco. 2011. Esparto wool as reinforcement in hybrid polyurethane composite foams. *Ind. Crops Prod.* 34:1641–1648.

Antunes, M., and J. I. Velasco. Multifunctional polymer foams with carbon nanoparticles. *Prog. Polym. Sci.*, under review.

Antunes, M., M. L. Maspoch, and J. I. Velasco. 2012. The multi-scalar effect of incorporating nanofillers and cellulosic-based reinforcements into polyurethane foams: Towards the development of low cost structural lightweight materials. Work presented in the *10th International Conference on Foam Materials & Technology*, Barcelona, Paper 36.

Antunes, M., J. I. Velasco, and V. Realinho. 2011. Polypropylene foams. Production, structure and properties. In *Advances in Materials Science Research*, Vol. 10, ed. M. C. Wythers, 121–152. New York: NovaScience Publishers.

Ashida, K. 1994. Thermosetting foams. In *Handbook of Plastic Foams*, ed. A.H. Landrock. Park Ridge, NJ: Noyes Publications.

Ashida, K. 2007. Polyurethane foams. In *Polyurethane and Related Foams. Chemistry and Technology*, ed. K. Ashida, 67–82. New York: CRC Press/Taylor & Francis.

Athanasopoulos, N., A. Baltopoulos, M. Matzakou, A. Vavouliotis, and V. Kostopoulos. 2012. Electrical conductivity of polyurethane/MWCNT nanocomposite foams. *Polym. Comp.* 33:1302–1312.

Backus, J. 2004. Rigid polyurethane foams. In *Polymeric Foams and Foam Technology*, ed. D. Klempner, and V. Sendijarevic, 121–140. Munich: Hanser Publishers.

Bandarian, M., A. Shojaei, and A. M. Rashidi. 2011. Thermal, mechanical and acoustic damping properties of flexible open-cell polyurethane/multi-walled carbon nanotube foams: Effect of surface functionality of nanotubes. *Polym. Int.* 60:475–482.

Banik, I. and M. M. Sain. 2008. Water blown soy polyol-based polyurethane foams of different rigidities. *J. Reinf. Plast.Comp.* 27:357–373.

Bernal, M. M., M. A. López-Manchado, and R. Verdejo. 2011. *In situ* foaming evolution of flexible polyurethane foam nanocomposites. *Macrom. Chem. Phys.* 212:971–979.

Bernal, M. M., I. Molenberg, S. Estravís, M. A. Rodríguez-Pérez, I. Huynen, M. A. López-Manchado, and R. Verdejo. 2012. Comparing the effect of carbon-based nanofillers on the physical properties of flexible polyurethane foams. *J. Mater. Sci.* 47:5673–5679.

Cao, X., L. J. Lee, T. Widya, and C. W. Macosko. 2005. Polyurethane/clay nanocomposites foams: Processing, structure and properties. *Polymer* 46:775–783.

Chen, X. G., X. M. Sang, G. X. Hou, and S. W. Yu. 2011. Preparation and mechanical properties of polyurethane/modified kaolin foam composites. *New Adv. Mater.* 197–198:1171–1175.

Chen, D. Q., J. Y. Yang, and G. H. Chen. 2010. The physical properties of polyurethane/graphite nanosheets/carbon black foaming conducting nanocomposites. *Comp. Part A-Appl. Sci. Manuf.* 41:1636–1638.

Christenson, E. M., J. M. Anderson, and A. Hittner. 2007. Biodegradation mechanisms of polyurethane elastomers. *Corros. Eng. Sci. Technol.* 42:312–323.

Chuayjuljit, S., A. Maungchareon, and O. Saravari. 2010. Preparation and properties of palm oil-based rigid polyurethane nanocomposite foams. *J. Reinf. Plast. Comp.* 29:218–225.

Dai, K., X. Ji, Z. D. Xiang, W. Q. Zhang, J. H. Tang, and Z. M. Li. 2012. Electrical properties of an ultralight conductive carbon nanotube/polymer composite foam upon compression. *Polym.-Plast. Technol. Eng.* 51:304–306.

Dolomanova, V., J. C. M. Rauhe, L. R. Jensen, R. Pyrz, and A. B. Timmons. 2011. Mechanical properties and morphology of nano-reinforced rigid PU foam. *J. Cell. Plast.* 47:81–93.

Eaves, D. 2004. Rigid polyurethane foams. In *Handbook of Polymer Foams*, ed. D. Eaves, 55–84. Shawbury: Rapra Technology Limited.

Fan, H., A. Tekeei, G. J. Suppes, and F. H. Hsieh. 2012. Properties of biobased rigid polyurethane foams reinforced with fillers: Microspheres and nanoclay. *Inter. J. Polym. Sci.* Vol. 2012:Article ID 474803.

Fan, X. and W. Xiao-Qing. 2009. Study on compress mechanical properties of reinforced polyurethane rigid foam. *Fiber Reinf. Plast./Compos.* 206:53–55.

Farkas, A., G. A. Mills, W. E. Erner, and J. B. Maerker. 1959. Triethylenediamine—A new bicyclic intermediate and catalyst for making polyurethane foam. *Ind. Eng. Chem.* 51:1299–1300.

Fromstein, J. D. and K. A. Woodhouse. 2002. Elastomeric biodegradable polyurethane blends for soft tissue applications. *J. Biomater. Sci. Polym. Ed.* 13:391.

Gournis, D., M. A. Karakassides, T. Bakas, N. Boukos, and D. Petridis. 2002. Catalytic synthesis of carbon nanotubes on clay minerals. *Carbon* 40:2641–2646.

Grenier, S., M. Sandig, D. W. Holdsworth, and K. Mequanint. 2009. Interactions of coronary artery smooth muscle cells with 3D porous polyurethane scaffolds. *J. Biomed. Mater. Res. A* 89:293–303.

Guo, A., D. Demydov, W. Zhang, and Z. S. Petrovic. 2002. Polyols and polyurethanes from hydroformylation of soybean oil. *J. Polym. Env.* 10:49–52.

Han, M. S., Y. H. Kim, S. J. Han, S. J. Choi, S. B. Kim, and W. N. Kim. 2008. Effects of a silane coupling agent on the exfoliation of organoclay layers in polyurethane/organoclay nanocomposite foams? *J. Appl. Polym. Sci.* 110:376–386.

Harikrishnan, G., T. U. Patro, and D. V. Khakhar. 2006. Polyurethane foam-clay nanocomposites: Nanoclays as cell openers. *Ind. Eng. Chem. Res.* 45:7126–7134.

Harikrishnan, G., S. N. Singh, E. Kiesel, and C. W. Macosko. 2010. Nanodispersions of carbon nanofiber for polyurethane foaming. *Polymer* 51:3349–3353.

Harrison, B. S. and A. Attala. 2007. Carbon nanotube applications for tissue engineering. *Biomaterials* 28:344–353.

Herrington, R., R. Broos, and P. Knaub. 2004. Flexible polyurethane foams. In *Polymeric Foams and Foam Technology*, ed. D. Klempner and V. Sendijarevic, 55–120. Munich: Hanser Publishers.

Hodlur, R. M. and M. K. Rabinal. 2012. Graphene based polyurethane material: As highly pressure sensitive composite. *Solid St. Phys. Pts. 1 and 2 AIP Conference Proceedings* 1447:1279–1280.

Housel, T. 2004. Flexible polyurethane foams. In *Handbook of Polymer Foams*, ed. D. Eaves, 85–122. Shawbury: Rapra Technology Limited.

Huang, M. N., Y. L. Wang, and Y. F. Luo. 2009. Biodegradable and bioactive porous polyurethanes scaffolds for bone tissue engineering. *J. Biomed. Sci. Eng.* 2:36–40.

Javni, I., K. Song, J. Lin, and Z. S. Petrovic. 2011. Structure and properties of flexible polyurethane foams with nano- and micro-fillers. *J. Cell. Plast.* 47:357–372.

Javni, I., W. Zhang, V. Karajkov, Z. S. Petrovic and V. Divjakovic. 2002. Effect of nano- and micro-silica fillers on polyurethane foam properties. *J. Cell. Plast.* 38:229–239.

Jell, G., R. Verdejo, L. Safinia, M. S. P. Shaffer, M. M. Stevens, and A. Bismarck. 2008. Carbon nanotube-enhanced polyurethane scaffolds fabricated by thermally induced phase separation. *J. Mater. Chem.* 18:1865–1872.

Jozwiak, A. B., C. M. Kielty, and R. A. Black. 2008. Surface functionalization of polyurethane for the immobilisation of bioactive moieties on tissue scaffolds. *J. Mater. Chem.* 18:2240–2248.

Kabir, M. E., M. C. Saha, and S. Jeelani. 2007. Effect of ultrasound sonication in carbon nanofibers/polyurethane foam composite. *Mater. Sci. Eng. A—Struct. Mater. Prop. Microstruct. Proc.* 459:111–116.

Kim, Y. H., S. J. Choi, J. M. Kim, M. S. Han, W. N. Kim, and K. T. Bang. 2007. Effects of organoclay on the thermal insulating properties of rigid polyurethane foams blown by environmentally friendly blowing agents. *Macrom. Res.* 15:676–681.

Kim, W., M. Han, J. Lee, and S. Choi. 2008. Preparation of polyurethane foam/clay nanocomposite insulating material involves add-mixing diisocyanate compound to mixture of silane coupling agent and organic clay, and dispersing organic clay into formed mixture. Patent KR2008063202-A, Korea.

Krishnamurthi, B., S. Bharadwaj-Somaskandan, and F. Shutov. 2001. Nano- and micro-fillers for polyurethane foams: Effect on density and mechanical properties. *Proceedings of Polyurethanes Expo*, Columbus, Ohio, Sept. 30-Oct. 3, 239–244.

Lee, L. J., C. C. Zeng, X. Cao, X. Han, J. Shen, and G. Xu. 2005. Polymer nanocomposite foams. *Comp. Sci. Tech.* 65:2344–2363.

Liang, K. and S. Q. Shi. 2010. Soy-based polyurethane foam reinforced with carbon nanotubes. *Adv. Des. Manuf. II* 419–420:477–480.

Liang, K. and S. Q. Shi. 2011. Nanoclay filled soy-based polyurethane foam. *J. Appl. Polym. Sci.* 119:1857–1863.

Li, M. K. S., P. Gao, P. L. Yue, and X. Hu. 2009. Synthesis of exfoliated CNT-metal-clay nanocomposite by chemical vapor deposition. *Sep. Purif. Technol.* 67:238–243.

Li, Y., H. F. Ren, and A. J. Ragauskas. 2010. Rigid polyurethane foam reinforced with cellulose whiskers: Synthesis and characterization. *Nano-Micro Lett.* 2:89–94.

Li, Y., H. F. Ren, and A. J. Ragauskas. 2011. Rigid polyurethane foam/cellulose whisker nanocomposites: Preparation, characterization, and properties. *J. Nanosci. Nanotechn.* 11:6904–6911.

Li, Z and Z. Xiang. 2009. Soft conductive polyurethane foam plastic for industrial applications comprises polyether glycol, vulcabond, steamed water, stannous octoate, triethylene diamine compounds, silicon oil and carbon nanotubes. Patent CN101250321-A, China.

Liu, M., Z. S. Petrovic, and Y. J. Xu. 2009. Bio-based polyurethane-clay nanocomposite foams: Synthesis and properties. *Arch. Multifun. Mater.* 1188:95–102.

Lorenzetti, A., D. Hrelja, S. Besco, M. Roso, and M. Modesti. 2010. Improvement of nanoclays dispersion through microwave processing in polyurethane rigid nanocomposite foams. *J. Appl. Polym. Sci.* 115:3667–3674.

Lu, M., K. T. Lau, J. Q. Qi, D. D. Zhao, Z. Wang, and H. L. Li. 2005. Novel nanocomposite of carbon nanotubes-nanoclay by direct growth of nanotubes on nanoclay surface. *J. Mater. Sci.* 40:3545–3548.

Luo, X. G., A. Mohanty, and M. Misra. 2012. Water-blown rigid biofoams from soy-based biopolyurethane and microcrystalline cellulose. *J. Am. Oil Chem. Soc.* 89:2057–2065.

MacDonald, R. A., B. F. Laurenzi, G. Viswanathan, P. M. Ajayan, and J. P. Stegemann. 2005. Collagen-carbon nanotubes composite materials as scaffolds in tissue engineering. *J. Biomed. Mater. Res.* 74A:489–495.

Madaleno, L., R. Pyrz, A. Crosky, L. R. Jensen, J. C. M. Rauhe, V. Dolomanova, A. M. Timmons, J. J. Pinto and J. Norman. 2013. Processing and characterization of polyurethane nanocomposite foam reinforced with montmorillonite-carbon nanotube hybrids. *Comp.: Part A* 44:1–7.

Madaleno, L., R. Pyrz, L. R. Jensen, J. J. C. Pinto, A. B. Lopes, V. Dolomanova, and J. Schjødt-Thomsen. 2012a. Synthesis and characterization of montmorillonite-carbon nanotubes hybrid fillers for nanocomposites. *Polym. & Polym. Comp.* 20:693–700.

Madaleno, L., R. Pyrz, L. R. Jensen, J. J. C. Pinto, A. B. Lopes, V. Dolomanova, J. Schjødt-Thomsen, and J. C. M. Rauhe. 2012b. Synthesis of clay-carbon nanotube hybrids: Growth of carbon nanotubes in different types of iron modified montmorillonite. *Comp. Sci. Technol.* 72:377–381.

Modesti, M., A. Lorenzetti, S. Besco, D. Hreja, S. Semenzato, R. Bertani, and R. A. Michelin. 2008. Synergism between flame retardant and modified layered silicate on thermal stability and fire behavior of polyurethane-nanocomposite foams. *Polym. Deg. Stab.* 93:2166–2171.

Mondal, P. and D. V. Khakhar. 2007. Rigid polyurethane-clay nanocomposite foams: Preparation and properties. *J. Appl. Polym. Sci.* 103:2802–2809.

Narine, S. S., X. Kong, L. Bouzidi, and P. Sporns. 2007. Physical properties of polyurethanes produced from polyols from seed oils: II. Foams. *J. Am. Oil Chem. Soc.* 84:65–72.

Nikje, M. M. A. and Z. M. Tehrani. 2011. The effects of functionality of the organifier on the physical properties of polyurethane rigid foam/organified nanosilica. *Design Mon. Polym.* 14:263–272.

Piticescu, R. M., L. M. Popescu, and T. Buruiana. 2012. Composites containing hydroxyapatite and polyurethane ionomers as bone substitution materials. *Dig. J. Nanomater. Biostruct.* 7:477–485.

Saha, M. C., M. E. Kabir, and S. Jeelani. 2008. Enhancement in thermal and mechanical properties of polyurethane foam infused with nanoparticles. *Mater. Sci. Eng.* A 479:213–222.

Sarier, N. and E. Onder. 2010. Organic modification of montmorillonite with low molecular weight polyethylene glycols and its use in polyurethane nanocomposite foams. *Thermo. Acta* 510:113–121.

Seal, B. L., T. C. Otero, and A. Panitch. 2001. Polymeric biomaterials for tissue and organ regeneration. *Mater. Sci. Eng. R: Reports* 34:147–230.

Seo, W. J., Y. T. Sung, S. J. Han, Y. H. Kim, O. H. Ryu, H. S. Lee, and W. N. Kim. 2006. Effects of ultrasound on the synthesis and properties of polyurethane foam/clay nanocomposites. *J. Appl. Polym. Sci.* 101:2879–2883.

Shaffer, M. S. P. and J. K. W. Sandler. 2007. Carbon nanotube/nanofibre polymer composites. In *Processing and Properties of Nanocomposites*, ed. S. G. Advani, 1–60. Singapore: World Scientific Publishing.

Singh, H. and A. K. Jain. 2009. Ignition, combustion, toxicity, and fire retardancy of polyurethane foams: A comprehensive review. *J. Appl. Polym. Sci.* 111:1115–1143.

Singh, V., D. Joung, L. Zhai, S. Das, S. I. Khondaker, and S. Seal. 2011. Graphene based materials: Past, present and future. *Prog. Mater. Sci.* 56:1178–1271.

Sung, C. H., K. S. Lee, K. S. Lee, S. M. Oh, J. H. Kim, M. S. Kim, and R. M. Jeong. 2007. Sound damping of a polyurethane foam nanocomposite. *Macromol. Res.* 15:443–448.

Valizadeh, M., M. Rezaei, and A. Eyvazzadeh. 2011. Effect of nanoclay on the mechanical and thermal properties of rigid polyurethane/organoclay nanocomposite foams blown with cyclo and normal pentane mixture. *Key Eng. Mater.* 471–472:584–589.

Velasco, J. I. and M. Antunes. PLA-based foams and related porous materials for tissue engineering. In *Encyclopaedia of Biomedical Polymers and Polymeric Biomaterials*, ed. M. Mishra (in press).

Verdejo, R., G. Jell, L. Safinia, A. Bismarck, M. M. Stevens, and M. S. P. Shaffer. 2009. Reactive polyurethane carbon nanotube foams and their interactions with osteoblasts. *J. Biomed. Mater. Res. Part A* 88A:65–73.

Verdejo, R., R. Stämpfli, M. Álvarez-Lainez, S. Mourad, M. A. Rodríguez-Pérez, P. A. Bruhwiler, and M. Shaffer. 2009. Enhanced acoustic damping in flexible polyurethane foams filled with carbon nanotubes. *Comp. Sci. Technol.* 69:1564–1569.

Wang, J., G. Si, W. Yi-Fei, W. Ping, W. Jian-Hua, and L. Yong-Hua. 2004. Preparation of nano silicon dioxide reinforced rigid polyurethane foam. *J. Natl. Univ. Defense Technol.* 4.

Widya, T. and C. W. Macosko. 2005. Nanoclay-modified rigid polyurethane foam. *J. Macrom. Sci., Part B: Phys.* 44:897–908.

Willemsen, A. 2012. An experimental characterization of the acoustically dissipative properties of light-weight nanocomposite polyurethane foams augmented with carbon nanotubes. *J. Acoust. Soc. Am.* 131:3271.

Xiang, Z. D., T. Chen, Z. M. Li, and X. C. Bian. 2009. Negative temperature coefficient of resistivity in lightweight conductive carbon nanotube/polymer composites. *Macromol. Mater. Eng.* 294:91–95.

Xie, H. and Z. Wang. 2005. In-site generating nano silicon dioxide and its effect as filler on properties of polyurethane rigid foam. *Plast. Sci. Technol.* 5:24–28.

Xu, Z. B., W. W. Kong, M. X. Zhou, and M. Peng. 2010. Effect of surface modification of montmorillonite on the properties of rigid polyurethane foam composites. *Chin. J. Polym. Sci.* 28:615–624.

Xu, X. B., Z. M. Li, L. Shi, X. C. Bian, and Z. D. Xiang. 2007. Ultralight conductive carbon-nanotube-polymer composite. *Small* 3:408–411.

Xu, Z., X. Tang, A. Gu, and Z. Fang. 2007. Novel preparation and mechanical properties of rigid polyurethane foam/organoclay nanocomposites. *J. Appl. Polym. Sci.* 106:439–447.

Yan, D. X., K. Dai, Z. D. Xiang, Z. M. Li, X. Ji, and W. Q. Zhang. 2011. Electrical conductivity and major mechanical and thermal properties of carbon nanotube-filled polyurethane foams. *J. Appl. Polym. Sci.* 120:3014-3019.

Yan, D. X., L. Xu, C. Chen, J. H. Tang, X. Ji, and Z. M. Li. 2012. Enhanced mechanical and thermal properties of rigid polyurethane foam composites containing graphene nanosheets and carbon nanotubes. *Polym. Int.* 61:1107–1114.

You, K. M., S. S. Park, C. S. Lee, J. M. Kim, G. P. Park, and W. N. Kim. 2011. Preparation and characterization of conductive carbon nanotube-polyurethane foam composites. *J. Mater. Sci.* 46:6850–6855.

Zammarano, M., R. H. Kramer, R. Harris, T. J. Ohlemiller, J. R. Shields, S. S. Rahatekar, S. Lacerda, and J. W. Gilman. 2008. Flammability reduction of flexible polyurethane foams via carbon nanofiber network formation. *Polym. Adv. Technol.* 19:588–595.

Zawadzak, E., M. Bil, J. Ryszkowska, S. N. Nazhat, J. Cho, O. Bretcanu, J. A. Roether, and A. R. Boccaccini. 2009. Polyurethane foams electrophoretically coated with carbon nanotubes for tissue engineering scaffolds. *Biomed. Mater.* 4:015008.

Zhang, J., B. A. Doll, E. J. Beckman, and J. O. Hollinger. 2003. A biodegradable polyurethane-ascorbic acid scaffold for bone tissue engineering. *J. Biomed. Mater. Res. A* 67:389–400.

Zhang, W. D., I. Y. Phang, and T. X. Liu. 2006. Growth of carbon nanotubes on clay: Unique nanostructured filler for high-performance polymer nanocomposites. *Adv. Mater.* 18:73–77.

Zhang, L., E. D. Yilmaz, J. Schjødt-Thomsen, J. C. Rauhe, and R. Pyrz. 2011. MWNT reinforced polyurethane foam: Processing, characterization and modelling of mechanical properties. *Comp. Sci. Tech.* 71:877–884.

# 7 The Use of Montmorillonite Clay in Polymer Nanocomposite Foams

*Priscila Anadão*

## CONTENTS

## 7.1 POLYMER/CLAY NANOCOMPOSITE TECHNOLOGY

The use of minerals as fillers in polymer materials was extensively employed until the late 1970s aiming to reduce final costs, since these fillers are heavier and cheaper than the polymers that had been used. As there was a vertiginous increase in petroleum prices during and after the 1973 and 1979 crises, in addition to the introduction of polypropylene on a commercial scale and the development of new materials containing mica, glass spheres and fibers, talc, and calcium carbonate, the ceramic raw material market as fillers was expanded and research regarding the interaction between polymers and fillers was initiated. This research was developed during the 1980s and the 1990s and contributed to raising minerals from a simple filler condition to a functional filler condition.

Polymer/clay nanocomposites were first prepared by Carter, Hendricks, and Bolley (1950), but only at the end of the 1980s was there the development of a great turning point in the polymer–clay nanocomposite technology by Toyota, using polyamide 6 and organophilic clays especially prepared for this nanocomposite (Kawasumi et al., 1989; Okada et al., 1988). They developed this nanocomposite to be applied in timing belt covers of Toyota vehicles, in collaboration with the UBE Industries, a Japanese polyamide 6 industry. This nanocomposite had only 5 wt% special clay, which sensibly improved the final material properties as compared with pure polyamide 6. The nanocomposite formation provided an increase of 40% in the rupture tension, 60% of the tensile modulus, and 126% in the flexion modulus, in addition to the increase in the thermal distortion temperature from 65 to 152°C as compared to the pure polymer.

Therefore, polymer/clay nanocomposites can be defined as a new class of composites with polymer matrices in which the dispersed phase is the silicate constituted by particles that have at least one of the dimensions at nanometer level. One of the components is the matrix, in which the particles of the second material are dispersed. The most used mineral particles in these nanocomposites are smectitic clays (montmorillonite, saponite, and hectorite), having their particles' lamellae morphology with sides at micrometer level and thickness around one nanometer (Alexandre and Dubois, 2000; Esteves, Barros-Timmons, and Trindade, 2004).

Many properties can be improved by the polymer/clay nanocomposite formation, for example, higher resistance toward high temperatures, UV radiation, high gas impermeability, low expansibility, and processing flexibility. Hence, they can be used in a wide range of applications. Furthermore, these nanocomposites are attractive since they can be processed by many types of processing, such as blending, compression, fusion, mixture, and polymerization, among others, satisfying premises for the production of the most different types of products, such as films, membranes, automotive parts, electronics, packaging materials, and so forth (Ke and Stroeve, 2005).

Several reviews about polymer/clay nanocomposites were carried out in recent years (Choudalakis and Gotsis, 2009; Mittal, 2009; Pavlidou and Papaspyrides, 2008; Ray and Okamoto, 2003b; Schmidt, Shah, and Giannelis, 2002; Yeh and Chang, 2008) and a large amount of research presents the use of montmorillonite (MMT) as clay. Then, considering the great attention paid to polymer/montmorillonite clay nanocomposites and to polymer foams, the presentation of research, which involves both topics, is opportune. In this chapter, nanocomposite morphologies, types of polymer–clay nanocomposite production, and modifications in polymer and montmorillonite structures, which allow them to be used in nanocomposite preparation, will be presented. The concepts involving foam production and its morphology will also be presented. Hence, having developed these concepts, works related to the use of montmorillonite in polymer nanocomposite foams will be discussed.

### 7.1.1 Polymer/Clay Nanocomposite Preparation Methods

Frequently, three preparation methods are widely used in polymer/clay nanocomposite technology. The first one is *in situ* polymerization. This method consists of the use of a monomer as a medium to the clay dispersion while favorable conditions are imposed to perform the polymerization between the clay layers. These layers present high surface energy and the monomer units are thus attracted to the inside of the galleries until equilibrium is reached. Polymerization can be initiated by heat or radiation, by the diffusion of an adequate initiator or a fixed catalyzer inside the layers before the filling step by the monomer. After that, polymerization reactions occur between the layers with lower polarity, dislocating the equilibrium and then aiming at the diffusion of new polar species between the layers.

Another method is solution dispersion where there is an exfoliation of silicate in single layers in a solvent medium in which the polymer or prepolymer is soluble. Since weak forces maintain united silicate layers, they can be easily dispersed in an adequate solvent by an increase in entropy caused by the disorganization of the layers, which exceeds the organizational entropy of the lamellas. The polymer is then sorbed into the delaminated layers and when the solvent is evaporated, or the mixture is precipitated, the layers are reunited, filled with the polymer.

The third method is fusion intercalation, a method developed by Vaia et al. in 1993 (Ma et al., 2012), which consists of a mixture of silicate with a thermoplastic polymer matrix in its melted state. Under these conditions and if the layer surface is sufficiently compatible with the chosen polymer, polymer chains can be dragged to the interlamelar space, forming the nanocomposite. The driving force in the melt intercalation process is the enthalpic contribution of the interactions between polymer and clay. The advantage of this technique is the nonuse of solvent (Ke and Stroeve, 2005; Souza, Pessan, and Rodolfo, 2006).

In addition to these three techniques, there is a less common method in which clay slurry is directly mixed with the polymer. The idea of this technique consists of increasing the basal spacing of the clay in the first stage of the process, by the polymerization of the monomers or clay suspension in water/solvents and hence, the polymer matrix is intercalated in the increased basal spacing (Kaneko, Torriani, and Yoshida, 2007). The use of supercritical carbon dioxide fluids and sol-gel technology can also be mentioned as preparation methods (Chuayjuljit, Maungchareon, and Saravari, 2010).

### 7.1.2 Polymer/Clay Nanocomposite Morphology

Three different types of nanocomposites are thermodynamically accepted according to the interfacial interactions forces between polymer matrix and silicate (Figure 7.1):

- *Intercalated nanocomposites*: In the intercalated nanocomposites, the insertion of polymer matrix in the silicate structure occurs in a regular way, crystallographically, by alternating polymer chains and silicate layer. The distance between each other varies from 1 to 4 nm, which is the distance in the range of the polymer chain length.

- *Flocculated nanocomposites*: These are the same as intercalated nanocomposites, except for the fact that some silicate layers are, sometimes, flocculated due to the interactions between the hydroxyl groups of the silicate.
- *Exfoliated nanocomposites*: In the exfoliated nanocomposites, the individual clay mineral layers are randomly separated in a continuous polymer matrix by an average distance which depends on the clay charge. Generally, the clay content in an exfoliated nanocomposite is much lower than in an intercalated nanocomposite (Ke and Stroeve, 2005; Ray and Okamoto, 2003b).

### 7.1.3  MONTMORILLONITE CLAY AND ITS ORGANOMODIFICATION

The mineral particles most used in these nanocomposites are the smectitic clays, for example, montmorillonite, saponite, and hectorite (Lee et al., 2005a; Liu et al., 2011). These clays belong to the 2:1 phylossilicate family.

Montmorillonite (MMT) clay can be studied by X-ray diffraction since it is composed of extremely small particles. The unitary layer of the montmorillonite consists of three sheets: one octahedral sheet composed of hydrargilite-brucite included between two tetrahedral sheets composed of silicon and oxygen. The tetrahedral edges of each silica layer form a common layer with one of the hydroxyl layers of the octahedral sheet, with oxygen atoms instead of hydroxyl groups (Papin, 1993). These layers are continuous in the a and b directions and are stacked in the c direction (Figure 7.2).

The clay thickness is around 1 nm and the side dimensions can vary from 30 nm to various micrometers, depending on the clay. The layer stacking by van der Waals and weak electrostatic forces originates the interlayer spaces or the galleries. In the layers, aluminium ions can be replaced by iron and magnesium ions, and as well, magnesium ions can be replaced by lithium ions, and so, the negative charge of the clay layers is neutralized by the alkaline and terrous-alkaline cations that are between these layers. Furthermore, there are water and other polar molecules which cause an expansion in the c direction, which can vary from 9.6 Å, when any polar molecule is between the unitary layers, up to complete separation in some cases (Brigatti, Galan, and Theng, 2006). That is, the expansion of the basal space leads to the occurrence of intercalated structures and the loss of registration of the ordered silicate layers produces exfoliated structures, which are responsible for the improvements observed in the properties of polymer/clay nanocomposites.

 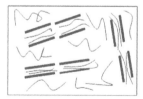 

Intercalated          Intercalated and Flocculated          Exfoliated

**FIGURE 7.1**  Illustration of the three types of thermodynamically accepted nanocomposites. (From Anadão P., 2011, In *Advances in Nanocomposite Technology*, ed. Hashim, A., 133–46. Rijeka: InTech.)

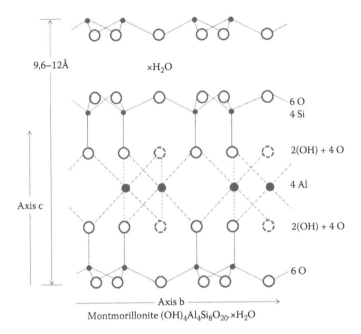

9,6–12Å    ×H₂O

Axis c

Axis b

Montmorillonite $(OH)_4Al_4Si_8O_{20}\cdot xH_2O$

6 O
4 Si

2(OH) + 4 O

4 Al

2(OH) + 4 O

6 O

**FIGURE 7.2**  Schematical representation of the montmorillonite structure. (From Anadão P., 2011, In *Advances in Nanocomposite Technology*, ed. Hashim, A., 133–46. Rijeka: InTech.)

This resulting surface charge is known as *cation exchange capacity* (CEC) and is expressed as mequiv/100 g. It should be highlighted that this charge varies according to the layer and is considered an average value in the whole crystal (Lee, Chen, and Hanna, 2008; Tomasko et al., 2003). The surface charge was described by Lagaly (1981) by considering the total elemental analysis and the dimension of the unit cell (Equation 7.1):

$$\text{Surface charge}: \frac{e^-}{nm^2} = \frac{\xi}{ab} \tag{7.1}$$

which $\xi$ is the layer charge (0.33 for MMT), $a$ and $b$ are the cell parameters (for MMT, 5.18 Å and 9.00 Å, respectively), whose surface charge value is 0.780 $e^-/nm^2$, which means that the average distance between exchange sites is 1.188 nm for MMT, by assuming that the cations are evenly distributed in a cubic array over the silicate surface and that half of the cations are located on one side of the platelet, and the other half are on the other side.

Frequently, in order to increase compatibility between the hydrophilic clay and the hydrophobic polymers, the clay surface is modified by organic surfactants.

The adsorption of organic materials by organic materials have been explored by humanity since 7000 B.C. (Lagaly, 1984), but the first research related to it appeared in the 1920s. Undoubtedly, Jordan (1949) contributed very much to this issue, as he presented the factors involved in the organoclay swelling and the extension of conversion of the clay from hydrophilic to hydrophobic.

The preparation of organoclays consists of reactions in which the water molecules in the interlayer spaces are displaced by polar molecules that can form complexes with the interlayer cations. Two methods are often used to prepare organoclays, which are the cation exchange and the solid-state reactions.

The first method consists of the exchange of the interlayer cation by quaternary alkylammonium cations in aqueous solution. And, the solid-state reaction performs this exchange without using solvent, which means that this method is environmentally friendly, since the dried clay is used in this process.

Different structures can be formed when the organoclay is produced. These structures are dependent mainly on the layer charge and on the chain length of the organic ion. Other factors can also influence the final structure, such as the geometry of the surface and the degree of exchange (Lagaly, 1981). Monolayer or bilayer structures can be produced, in addition to an inclined paraffin-type structure and pseudotrimolecular layers (Figure 7.3).

As mentioned before, the most common cationic surfactants used in the organomodification are the quaternary alkylammonium salts, which are generally synthesized by complete alkylation of ammonia or amines. Figure 7.4 presents some examples of these alkylammonium salts.

The type of quaternary alkylammonium salt influences the affinity between polymer chains and the clay mineral platelets. As an example, nonpolar polymers, such as polypropylene and polyethylene, are more compatible with organoclays prepared with dimethylammonium halides, while polar polymers present more affinity with the organoclays containing alkyl benzyl dimethylammonium halides and alkyl hydroxyethylammonium halides.

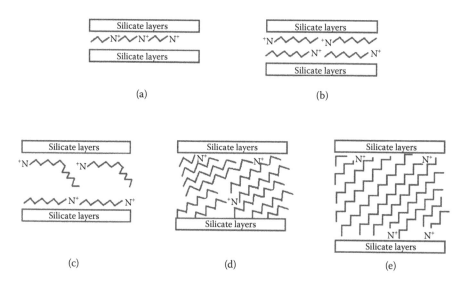

**FIGURE 7.3**  Organoclay structures: (a) monolayer, (b), bilayer, (c) pseudotrimolecular layer, and (d) and (e) paraffin-type arrangement of the organic molecules within the clay mineral layers. (From Bergaya F. and Lagaly G., 2001, *Applied Clay Science* 19, 1–3.)

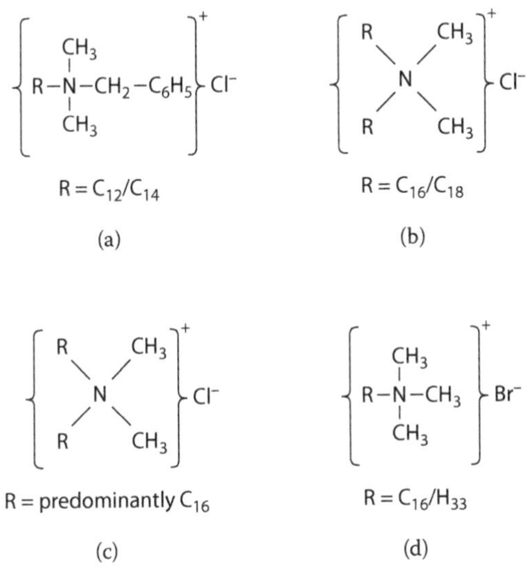

**FIGURE 7.4** Molecular structure of the ammonium quaternary salts: (a) Dodigen®, (b) Praepagen®, (c) Genamin®, (d) Cetremide®. (From Barbosa R. et al., 2006, *Polímeros: Ciência e Tecnologia* 16, 246–51.)

In addition to ammonium quaternary salts, other surfactants are also mentioned in the literature as an organomodifier in organoclays. Nonionic surfactants, which are linear alcohol ethoxylate, with low toxicity and potential for biodegradation are used to prepare organomontmorillonite (Shen, 2001). Furthermore, clays can also be modified with biomolecules, such as proteins, enzymes, amino acids, peptides, and so forth.

### 7.1.4 POLYMER GRAPHITIZATION

In addition to clay organomodification, changes can be performed in the polymer in order to increase polymer–clay compatibility. A compatibilizing agent, generally a polymer, can be used to offer a chemically compatible nature with the polymer and the clay. By a treatment, such as the graphitization of a chemical element that has reactive groups, or copolymerization with another polymer, which also has reactive groups, polymer becomes more compatible with the clay. Hence, amounts of the modified polymer are mixed with the polymer without modification and the clay to prepare the nanocomposites.

Aiming to increase compatibility between polymer and clay, parameters such as molecular mass, type and content of functional groups, compatibilizing agent/clay proportion, and processing method should be taken into account. Maleic anhydride is the organic substance most used to compatibilize polymer, especially with polyethylene and polypropylene, since the polar character of maleic anhydride results in favorable interactions, creating a special affinity with the silicate surfaces (Fu and Naguib, 2006; Strauss and D'Souza, 2004).

## 7.2   FOAMS

Polymer foams, or porous polymeric materials, can be defined as dispersions of a gas in a liquid, which, once solidified, consist of individual cells (pores) and walls that form a skeletal structure. Foam production is an alternative to decrease the amount of resin used, as foams keep acceptable mechanical properties and can be produced as several types of morphology. Hence, they are used in many applications, which include insulating, packaging, textile, agriculture, construction, and automotive industries, due to their low weight, excellent strength-to-weight ratio, and good thermal and sound insulation.

Foams are often classified by three main parameters, which are across-cell nominal diameter or cell size, BET (Brunnauer–Emmett–Teller) surface area ($m^2/g$), and cell density (number of cells per unit volume [cells/$cm^3$]). According to the cell size, polymer foams can be classified into:

- Macrocellular foams (>100 µm), which are very attractive mostly because of their lower cost per volume unit compared to unfoamed materials, but also, for their sound or heat insulating properties, cushioning ability, and so forth. However, the foam mechanical properties are generally lower than plain material ones, which limit their range of applications.
- Microcellular foams (1–100 µm), which were first developed by Martini (1981) about 30 years ago, present superior mechanical properties in terms of their impact strength, toughness, and fatigue life over conventional foams and can also provide better thermal insulation and acoustic properties.
- Ultramicrocellular foams (0.1–1 µm).
- Nanocellular foams (0.1–100 nm).

Regarding the rigidity of the skeletal structure, the foams can also be categorized as rigid (largely employed in building insulation, appliances, transportation, furniture, etc.) and flexible foams. Still, according to the nature of the cells, the foams can also be defined as either closed or open cell foams. The closed cell foams present cells isolated from each other and the cavities are surrounded by complete cell walls. Open cell foams are composed of broken cell walls and the structure is constituted by ribs and struts. Closed cell foams present lower permeability than open cell foams, which leads to better insulation properties.

The mechanical and thermal properties of the foams depend mainly on the relative density, which is the foam density divided by the solid density. The linear elastic properties of foams can be described as a function of relative density by Equation (7.2):

$$\frac{Foam\ property}{Solid\ property} = C\left(\frac{\rho f}{\rho s}\right)^{n} \tag{7.2}$$

where $C$ is the property of the polymer matrix which includes all the geometric constants of proportionality, $n$ represents the deformation mode of the struts that make up the foam (tensile or compressive) and is characterized by values between

1 and 427, and $\rho$ is the density, for which the $s$ and $f$ refer to the solid and the foam structures, respectively. $C$ and $n$ are dependent on the microstructure of the foam including the cell type (open or closed), geometrical arrangement of cells, cell size, and angle of intersection (Abu-Zahra and Alian, 2010; Chang et al., 2011; Ibeh and Bubacz, 2008; Istrate and Chen, 2011; Mitsunaga et al., 2003; Modesti, Lorenzetti, and Besco, 2007; Taki et al., 2004; Urbanczyk et al., 2010a, 2010b).

### 7.2.1 FOAM PREPARATION METHODS

Different methods can be used to prepare foams. The most common is extrusion, in which a blending often composed of a physical blowing agent and molten polymer is formed in order to produce a homogenized high-pressure solution for instantaneous free expansion at the extrusion exit, where the foam will be generated. This process comprises four steps, which are: (a) the dissolution of the foaming agent into a polymer melt at elevated pressure, (b) the nucleation of bubbles in a supersaturated solution of foaming agent in a molten polymer, (c) the growth of bubbles in a molten polymer, and (d) the stabilization of the cellular structure by lowering the temperature below the melting point or the glass transition temperature of polymer (Zandi, Rezaei, and Kasiri, 2011).

Another method is the compression molding process, which consists of a previously extrusion-compounded material that is foamed inside a hot-plate press or a modified oven by simultaneously applying temperature and pressure to gradually foam the material. Generally, a chemical blowing agent is employed and the process can be carried out in a single or a two-step foaming process. In this process, the cellular structure can be controlled by regulating the decomposition temperature of the chemical foaming agent and the melt viscosity of the polymer. If the temperature is high, the decomposition rate will be fast and the melt strength of the polymer will be low, resulting in coalescence and cell rupture. On the other hand, if the decomposition temperature is low, the decomposition rate will decrease—which will require higher foaming times—and the melt viscosity and strength of the base polymer will be much higher, restraining cell growth and resulting in only partially foamed products (Antunes et al., 2009).

The injection molding process can also be mentioned as a foaming preparation method. And, less frequently, there is also the high internal phase emulsion (HIPE), which consists of an emulsion with a large volume of internal liquid phase, at least 74% (by volume), which is dispersed in a small volume of monomeric continuous phase. An open-cellular solid porous material will be provided by the polymerization of the HIPE template and the subsequent removal of the locked internal phase from the solidified foam emulsion will produce the poly(HIPE). The cells or voids in the solid foam are characterized by being interconnected with small intercellular pores. These voids usually have a mean diameter between 5 and 100 µm, while the pore average sizes vary between 1 and 10 µm (Moghbeli and Shahabi, 2011).

Furthermore, the foaming process can be performed in a batch or continuous system. Batch foaming is characterized by being simple and easy to control. The material is foamed inside an autoclave reactor by a high-pressure gas dissolution process and then bubble nucleation and growth are induced by quick pressure release (pressure quench) or temperature increase (Antunes et al., 2009; Guo, Lee, and Tomasko, 2008).

Regarding the types of foaming agents, there are two types, which are the physical and the chemical blowing agents. The chemical blowing agents consist of reactive species that produce gases in the foaming process. The reactive species decompose inside the press at a given temperature and single temperature (to produce foams higher than $100 \text{ kg/m}^3$) and two-step foaming process (foams lower than $100 \text{ kg/m}^3$). Most thermoplastic foams are synthesized via a two-step process and the nanocomposites as well (Ibeh and Bubacz, 2008).

The physical foaming agents are substances that gasify under foaming conditions. Generally, the physical foaming processes can be performed in two ways: batchwise or continuously. The batchwise process can be conducted by saturating the polymer with gas (e.g., $CO_2$ and $N_2$) at room temperature, subsequently releasing the gas and foaming the polymer by heating it to a temperature above its softening point in a high temperature bath. However, during the transfer of gas-saturated materials to the high temperature environment, diffusion would occur inevitably, which leads to a lower cell density and expansion ratio. Also, in this process, the saturation time is very long, occurring from hours to days depending on diffusivity, which limits the production. The other way in which a batchwise process can be performed is by saturating polymer with gas at a relatively high temperature and pressure (in supercritical conditions) followed by rapid depressurization to atmospheric pressure. This process takes advantage of the depression of polymer softening temperature induced by the presence of gas (Khorasani et al., 2010).

Several gases can be used in polymer foaming. Due to the prohibition of the chlorofluorocarbons and hydrochlorofluorocarbons, which are ozone-depleting substances, other gases should be employed in the foaming processes such as the hydrofluorocarbons, which are greenhouse gases but expensive, and the hydrocarbons, which present high flammability and a volatile nature. Hence, inert gases, such as $H_2O$, $CO_2$, and $N_2$, are preferable due to their better solubility, nonflammability, and nontoxicity, in addition to being inexpensive and environmentally friendly.

The most used blowing agent is $CO_2$, which presents as advantageous because it is nontoxic, nonflammable, noncorrosive, abundant, inexpensive, commercially available in high purity, and has readily accessible supercritical conditions ($T_c = 31.1°C$ and $P_c = 7.37 \text{ MPa}$). Moreover, it has substantial solubility in amorphous and semicrystalline polymers and it can be used in a wide range of preparation methods, for example, in the liquid/melt state by extrusion, injection molding or compression molding, or in solid state, in which gas is forced into a solid polymer followed by depressurization. Finally, as no residue is left, it can be simply removed from the polymer matrix by a simple depressurization step (Liao, Zhang, and He, 2012).

By using $CO_2$, the gas is diffused into the polymer matrix, forming a polymer/$CO_2$ solution in which the density, the surface tension, and the viscosity of the polymer are reduced and the mobility of polymer chains is increased. As the system reaches equilibrium, the phase separation is induced by either reducing pressure (pressure quench method) or increasing temperature (temperature soak method), which produces nuclei that will grow and form the porous structure. It is important to mention that the $CO_2$ sorption level is dependent on the temperature, pressure, and intermolecular interactions between $CO_2$ and polymer (Han et al., 2003; Urbanczyk et al., 2010b). Hence, the factors that influence the nucleation rate are the $CO_2$ diffusivity, gas concentration in polymer, interfacial tension in the polymer/gas mixture, temperature, pressure drop rate, and the degree of supersaturation.

A different foaming was also proposed by Goel and Beckman which uses supercritical $CO_2$, obtained above the critical pressure and temperature. Its advantage when used as the physical blowing agent is that it will depress the glass transition temperature of the polymer and will prevent the vapor/liquid boundary, which can damage the cellular structure. The critical temperature of the $CO_2$ is 304.15 K and its critical pressure is 7.38 MPa, making it a good solvent. When used in foaming processes, a polymer sample is saturated with supercritical $CO_2$ and the microcellular foam is produced by a fast depressurization (Wee, Seong, and Youn, 2004).

### 7.2.2 POLYMER/CLAY NANOCOMPOSITE FOAMS

A successful manufacture of microcellular foams requires high-pressure drop, fast cooling rate, and less bubble coalescence during foaming to control the cell size and cell density (Zhu et al., 2010). Moreover, polymer foams represent a group of lightweight materials whose applications are limited due to their inferior mechanical strength, poor surface quality, and low thermal and dimensional stability. Considering this problem, the use of new technologies has come to improve foam properties.

The preparation of nanocomposite foams has been shown to improve cell morphology, making them smaller and more isotropical and resulting in enhanced thermomechanical properties with respect to the neat polymer foams. A small amount of well-dispersed nanoparticles in the polymer may serve as nucleation sites to facilitate the bubble nucleation process and thus, the reduction of the cell size is explained as more bubbles start to nucleate concurrently; there is a smaller amount of gas available for bubble growth in the presence of the nanoparticles, leading to a reduction of cell size. Therefore, the size, shape, and distribution of the nanoparticles can affect the nucleation efficiency. The amount and distribution of the nucleation agents also determine foam quality. The extremely small dimensions and large surface area of nanoparticles provide much more intimate contact between the particles, polymer matrix, and gas (Chang et al., 2011; Jiang et al., 2009; Keramati et al., 2012; Zeng et al., 2003).

As the polymer/clay nanocomposite technology presents several final material improvements on thermal, mechanical, chemical, and barrier properties, the use of clay nanoparticles was also thought to improve material and foam properties at the same time. Hence, the development of nanocomposite foams is one of the latest evolutionary technologies of polymeric foam through pioneering efforts by Okamoto and colleagues, who prepared poly(L-lactide)/clay nanocomposite foams in a batch process by using supercritical $CO_2$ as a physical foaming agent (Ema et al., 2006).

The clay nanoparticles act as nucleation agents for bubble generation in foams using $CO_2$ as a physical foaming agent, via batch or direct extrusion. In both cases, clay reduces the cell size and increases cell density. Generally, the cell density increases linearly with clay concentration at low clay concentrations up to 10 wt%. Moreover, cell nucleation can be improved if an exfoliated-type of structure is achieved by the clay particles, with finer particles reducing the nucleation energy for the growth of the gaseous phase. Hence, clay particles act as a reinforcing agent, expanding the range of properties and creating mechanically improved foams for structural applications (Ibeh and Bubacz, 2008).

### 7.2.3 A BRIEF THEORY CONCERNING THE NUCLEATION AGENT

The nucleation agent used in foaming preparation is very important since it determines the final foaming properties. The number and the size of the bubbles as well as the foam cell density are determined by the concentration of the nucleation agent. Its distribution is also important since a nonuniform distribution of the nucleation agent leads to a nonuniform cell size distribution result, characterized by a higher number of cells in the nucleation agent rich area and a lower number of cells in the areas with a lower content of nucleation agent.

There are two types of nucleation: the homogeneous one and the heterogeneous one. The classic nucleation theory is the approach used to describe bubble nucleation in polymeric foams, although there is a discrepancy attributed to the intervening heterogeneous nucleation.

The steady-state nucleation rate, $N_0$, is given by Equation (7.3):

$$N_0 = \frac{C_0 f_0 exp\left(-\Delta G_{crit}\right)}{k_B T} \tag{7.3}$$

where $\Delta G_{crit}$, $k_B$, and $T$ are the free energy of critical nucleus formation, the Boltzmann constant, and the absolute temperature, respectively. $C_0$ is the number of gas molecules dissolved per volume of the primary phase and $f_0$ is a kinetic pre-exponential factor that is believed to be weakly dependent upon temperature.

The incorporation of clay induces heterogeneous nucleation because of a lower activation energy barrier compared with the homogeneous nucleation, although the competition between the homogeneous and the heterogeneous is no longer discernible. Then, the nucleation rate in the heterogeneous nucleation, $N_1$, is expressed by Equation (7.4):

$$N_1 = \frac{C_1 f_1 exp\left(-\Delta G_{crit}^{het}\right)}{k_B T} \tag{7.4}$$

where $f_1$ is the frequency factor of gas molecules joining the nucleus and $C_1$ is the concentration of heterogeneous nucleation sites. The $\Delta G_{crit}^{het}$ is dependent on the contact angle between the gas, polymer, and particle surface and from the particle surface curvature (Ibeh and Bubacz, 2008).

## 7.3 EXAMPLES OF POLYMER/MONTMORILLONITE NANOCOMPOSITE FOAMS

It is well-known that montmorillonite (MMT) clay is the most used clay in polymer/clay nanocomposite foams. Hence, it is suitable to present the effect of adding montmorillonite clay in the preparation of polymer nanocomposite foams and its effect as nucleation agent in the final foam properties. Next, scientific advances related to the use of montmorillonite are presented, these studies being divided into general purpose plastics, engineering plastics, elastomeric polymers, biodegradable polymers, and natural polymers, and correlating them with the production method, final morphology, and resulting properties.

### 7.3.1 General Purpose Plastic/Montmorillonite Nanocomposite Foams

General purpose plastics are characterized for being used in low cost applications due to the ease of processing and low level of mechanical exigency. One way to add value to these commodities is nanocomposite formation, aiming to improve their properties (Anadão et al., 2011).

A large number of papers focus on the use of polystyrene (PS) in the production of nanocomposite foams (Guo, Lee, and Tomasko, 2008; NGO et al., 2010; Wee, Seong, and Youn, 2004; Zhu et al., 2010). One of these studies showed that by combining nanocomposites and the extrusion foaming process, the cell structure could be designed and controlled in microcellular foams. Polystyrene/organomodified MMT (Cloisite 20A) nanocomposites were prepared by mechanical blending and *in situ* polymerization. By using an extrusion foaming process with $CO_2$ as foaming agent at a screw rotation speed of 10 rpm and a die temperature at 200°C, it was found that the cell density increased linearly with the clay content increase or pressure drop rate. The MMT addition increased the nucleation sites and then the nucleation rate, according to the heterogeneous nucleation theory. A better dispersion of the clay particles provided more nucleation sites than the less well-dispersed clay particles, since the exfoliated nanocomposite presented the highest cell density ($1.5 \times 10^9$ cells/cm$^3$) and the smallest cell size (4.9 μm). The nanocomposite foams also showed higher tensile modulus, improved fire retardance, and better barrier properties in comparison with the pristine polystyrene foams (Han et al., 2003; Zeng et al., 2003).

Suspension polymerization of water-in-oil inverse emulsion was also employed to produce water expandable PS/sodium MMT nanocomposite foams. The use of water as a carrier allowed the MMT incorporation into the polymer beads, which helped to trap more water in the beads during synthesis and to reduce the water loss during storage. This presence of water enlarged the cell size (~100 μm) and decreased the foam density (< 0.05 g/cm$^3$) (Shen, Cao, and Lee, 2006).

Another polymer largely used in the preparation of MMT nanocomposite foams is the high-density polyethylene (HDPE) (Jo and Naguib, 2007a,b,c; Lee et al., 2005b). Microcellular HDPE-MMT nanocomposite foams were prepared via a batch high temperature process using supercritical $N_2$. The volume expansion and the cell density were improved by MMT use since it increased the cell nucleation and $N_2$ sorption. The MMT dispersion affected the microcellular morphology, and its better dispersion increased the number of nucleation sites available. The crystalline morphology also played an important role in the cell growth mechanism of this process since the solid-state nucleation was followed by a cell growth in a softened system strengthened by its degree of crystallinity and melting point (Khorasani et al., 2010).

Poly(methyl methacrylate) (PMMA) was also used in MMT nanocomposite foam preparation (Fu and Naguib, 2006; Weickmann et al., 2010). Organomodified MMT was used in its preparation by *in situ* bulk polymerization and batch foaming process with $N_2$. The MMT acted as a heterogeneous nucleation agent in polymer matrix that reduced cell size and increased cell density. Moreover, MMT addition decreased dielectric constant and slightly increased the thermal conductivity as well as the mechanical strength due to the well-dispersed clay mineral platelets (Yeh et al., 2009).

Poly(styrene-co-acrylonitrile) (SAN)/MMT nanocomposite foams were prepared with the supercritical $CO_2$ technique (Urbanczyk et al., 2010b). The influence of two batch foaming processes was investigated and the use of a one-step foaming process at 100°C, also called *depressurization foaming* technique, showed that clay had little influence on the foam density, but doubled the cell density. The temperature also affected clay role as cell density was slight, increased at 60°C, and was largely increased at 40°C, showing the importance of the foaming condition influence when dealing with heterogeneous nucleation (Urbanczyk et al., 2010a).

Furthermore, as rigid poly(vinyl chloride) (PVC) foam is becoming increasingly popular in the building materials industry and sometimes its use is limited due to its inferior strength, poor surface quality, and low thermal and dimensional stability, the preparation of nanocomposite foams is welcome. Thus, two types of montmorillonite were used to prepare nanocomposite foams: calcium montmorillonite and montmorillonite modified with a quaternary ammonium salt and the foams were prepared by extrusion with the chemical blowing agent called *azodicarbonimide*. As both MMT content decreased, the cell diameter decreased whereas the cell number increased since the nucleation rate was enhanced by significantly decreasing the activation barrier, resulting in a large number of cells at a controlled size. The effect of both MMT on decreasing the foam density was more significant at lower levels of blowing agent. Still, both MMTs produced exfoliated nanocomposites and as it is known, the more dispersed the clay platelets are, the more efficient the nucleation is (Abu-Zahra and Alian, 2010).

### 7.3.2 ENGINEERING PLASTIC/MONTMORILLONITE NANOCOMPOSITE FOAMS

As an example of an engineering plastic used in the foam production, polyimide (PI) can be cited because of their unique properties, such as good thermal and acoustic insulation, high thermal stability, and excellent fire resistance; and thus, PI foams have been widely used in many fields, as for example, spacecraft, aircraft, marine, and high-speed train. As technologies advance, new possible applications demand better properties and so, new PI foams with improved properties will be developed. Hence, lightweight PI/MMT nanocomposite foams were prepared by solid blending, which means that MMT and PI were blended in the solid state by super-high speed mechanical shearing to conduct the MMT dispersion. Exfoliated nanocomposites were obtained as a result of the good MMT dispersion.

The average cell size of the nanocomposite foams decreased with the increase of the MMT content due to the nucleation effect of the MMT. As mentioned before, nucleation efficiency was enhanced due to the extremely fine dimensions and large surface area of the MMT that provided a more intimate contact between the PI chains and the gas. Regarding the mechanical properties, with the MMT increase to 7 wt%, the partial MMT agglomeration led to a marked decrease in the reduced compressive strength. Moreover, storage modulus increased to 5 wt% MMT and then decreased continuously with the increase of MMT content to 7 wt% as a result from the strong interaction between the MMT and the PI matrix to 5 wt%. Still, the thermal stability was improved by the MMT incorporation, denoted by an increase of the temperature for 10 wt% mass loss, as a result of the thermal isolation effect of the MMT. Finally, the nanoscopic-confinement effects in the randomly exfoliated and intercalated layer structures were responsible for the decrease in the dielectric constants of the nanocomposite foams (Pan, Zhan, and Wang, 2011).

### 7.3.3 ELASTOMERIC POLYMER/MONTMORILLONITE NANOCOMPOSITE FOAMS

Elastomeric polymers are widely used in applications such as foamed sheets, automotive parts, durable goods, impact-modifiers in engineering plastics, and wire and cable owing to their outstanding heat, ozone, and weather resistance.

Ethylene-propylene-diene terpolymer (EPDM)/organomodified MMT (Cloisite 20A) nanocomposite foams were prepared by melt blending followed by foaming with $CO_2$ in an autoclave. Hybrid morphology, that is, exfoliated/intercalated nanocomposites were obtained. By increasing the clay content, the tensile modulus and tensile strength were increased due to the interfacial action between polymer and clay nanoparticles. Cell size was decreased by half (from 12.0 μm to 6.2 μm) and the cell density was increased to $2.4 \times 10^{10}$ cell/cm³, about 5.7 times larger, in comparison with the pure EPDM foam. These values were attributed to the larger effective nucleation sites provided by the fine dispersion of the clay particles in the nanocomposite (Chang et al., 2011).

Polyethylene-octene elastomer (POE) was also used to prepare Cloisite 20A-containing nanocomposite foams with poly[ethylene-*co*-(methyl acrylate)-*co*-(glycidyl methacrylate)] as a compatibilizing agent by melt blending followed by foaming with $CO_2$ in an autoclave. Intercalated nanocomposites were produced. Moreover, the tensile modulus of POE was increased by the stiffening effect of the clay particles and the dynamic storage moduli were enhanced due to the huge interfacial action between the polymer and the clay nanoparticles. The foams exhibited a closed-cell structure and the increase of the cell density was explained by the much larger effective nucleation sites in comparison with the pure polymer foam. The effective increase in the modulus due to the presence of clay nanoparticles also restrained the cell growth and their coalescence, which resulted in a cell size reduction (Chang, Lee, and Bae, 2006).

Still, open cellular elastomeric nanocomposite foams were obtained by polymerization of water-in-oil high internal phase emulsions containing 2-ethylhexylacrylate, styrene, divinylbenzene, and organomodified montmorillonite (Cloisite 15A and Cloisite 30B). Cloisite 15A produced foams with pore sizes smaller than the foams with Cloisite 30B due to the type of organoclay, which influenced the phase separation behavior within the polymerizing organic phase. Intercalated nanocomposites were obtained, although the mechanical properties were not improved (Moghbeli and Shahabi, 2011).

### 7.3.4 Biodegradable Polymer/Montmorillonite Nanocomposite Foams

As the plastics produced from fossil fuels, when discarded into the environment, end up as waste that cannot degrade spontaneously, there is an urgent need for the development of biodegradable polymers. However, their low thermal stability and mechanical properties usually limit their application. Therefore, nanocomposite preparation is an option to overcome these problems.

Poly($\varepsilon$-caprolactone) (PCL)/organomodified MMT nanocomposite foams prepared by chemical foaming with azodiformamide were reported in the literature. In comparison with the pristine foam, the nanocomposites presented an enhanced compressive property. By increasing clay content, the pore size decreased and the pore wall thickness remained almost the same at low clay contents. However, at high clay contents, both parameters increased due to the change in the melt viscosity of nanocomposites and the heterogeneous nucleation behavior of the clay at low contents (Liu et al., 2009).

Another biodegradable polymer used to prepare nanocomposite foams is poly(lactic acid) (PLA), which is used in service ware, grocery, waste-composting bags, mulch films, controlled release matrices for fertilizers, pesticides, and herbicides (Di et al., 2005; Lee and Hann, 2008; Lee et al., 2008; Ma et al., 2012; Ray and Okamoto, 2003a). The PLA nanocomposite foams prepared with two types of organomodified MMT via a batch process in an autoclave with $CO_2$ can be mentioned. Intercalated nanocomposites were obtained. At low foaming temperatures, the nanocomposite foams showed smaller cell size and larger cell density in comparison with the neat PLA foam, which suggested that the dispersed clay nanoparticles acted as nucleation sites for cell formation and these properties were a result of heterogeneous nucleation since it had a lower activation energy barrier as compared with the homogeneous nucleation. This conclusion was evidenced by the characterization of the interfacial tension between bubble and matrix. Finally, the foams obtained had pores in the range of microcellular (30 μm) to nanocellular (200 nm) (Ema, Ikeya, and Okamoto, 2006).

### 7.3.5 Natural Polymer/Montmorillonite Nanocomposite Foams

A different technique was used to prepare chitosan/xanthan gum/sodium montmorillonite nanocomposite foams. Aqueous colloidal suspension was prepared and a gel structure was formed as a consequence of freezing, called *cryogel*, which is a gel formed due to the concentration increase of the substrates caused by the ice formation during freezing. This gel was dried, producing a macroporous foam. Exfoliated nanocomposites were produced and the influence of the freezing method was studied in terms of foam morphology. It was found that, in the case of the contact freezing, rapid freezing (–2°C/min) produced randomly aligned pores in comparison with the foam produced by slow freezing (–0.25°C/min) whereas the mean pore size for rapid freezing and slowing freezing were 40 μm and 68 μm, respectively. Again, the use of MMT improved the hardness of the prepared foams (Liu, Han, and Dong, 2009).

## 7.4   FUTURE PERSPECTIVES

A great number of studies focuses their attention on general purpose plastic/montmorillonite nanocomposite foams, although great attention has also been given to biodegradable polymer nanocomposite foams. Moreover, it is important to mention that a great deal of research has used organomodified montmorillonite in these nanocomposite foams as this type of montmorillonite presents higher compatibility with the polymer chains and thus the clay particle dispersion is high, promoting a large number of nucleation sites, and hence, a higher cell density and a smaller cell size.

The use of montmorillonite clay was especially beneficial for the production of microcellular foams, which are considered for structural applications since they are lightweight and a high-strength material. By adding montmorillonite, foam and material properties were improved, which expands the range of applications.

However, the research presented refers to laboratory-scale production and in aiming to turn these materials toward commercial availablility, several developments must be performed, for example, the production of organomodified montmorillonite in large scale and low cost. Moreover, new types of organomodified montmorillonite should be investigated, for example, with nonionic surfactants, in order to obtain nanocomposites, which are completely exfoliated since it is still a challenge in the polymer/clay nanocomposite technology. By developing new types of organomodified montmorillonite, nucleation and growth of the bubbles can also be controlled rigorously. If these challenges are overcome, the use of montmorillonite in polymer nanocomposite foams can be a successful way to produce foams in a wide range of pore sizes that can be used in several types of applications.

## REFERENCES

Abu-Zahra, N. H. and Alian, A. M., 2010. Density and cell morphology of rigid foam PVC-clay nanocomposites. *Polymer-Plastics Technology and Engineering* 49:237–43.

Alexandre, M. and Dubois, P., 2000. Polymer-layered silicate nanocomposites: Preparation, properties and uses of a new class of materials. *Materials Science & Engineering R* 28: 1–63.

Anadão, P., 2011. Clay-containing polysulfone nanocomposites. In *Advances in Nanocomposite Technology*, ed. A. Hashim, 133–46. Rijeka: InTech.

Anadão P., 2012. Polymer/clay nanocomposites: Concepts, researches, applications and trends for the future. In *Nanocomposites: New Trends and Developments*, ed. F. Ebrahimi, 1–16. Rijeka: InTech.

Anadão, P., Wiebeck, H., and Díaz, F. R. V., 2011. Panorama da pesquisa acadêmica brasileira em nanocompósitos polímero/argila e tendências para o futuro. *Polímeros: Ciência e Tecnologia* 21: 443–52.

Antunes, M., Velasco, J. I., Realinho, V., and Solórzano, E., 2009. Study of the cellular structure heterogeneity and anisotropy of polypropylene and polypropylene nanocomposite foams. *Polymer Engineering and Science* 49: 2400–13.

Barbosa, R., Araújo, E. M., Maia, L.F., Pereira, O.D., Mélo, T.J.A., and Ito, E.N., 2006. Morfologia de nanocompósitos de polietileno e poliamida–6 contendo argila nacional. *Polímeros: Ciência e Tecnologia* 16: 246–51.

Bergaya, F. and Lagaly, G., 2001. Surface modifications of clay minerals. *Applied Clay Science* 19: 1–3.

Brigatti, M. F., Galan, E., and Theng, B. K. G., 2006. Structures and mineralogy of clay minerals. In *Handbook of Clay Science*, ed. F. Bergaya, B. K. G. Theng, and G. Lagaly, 19–86, Amsterdam: Elsevier.

Carter, L. W., Hendricks, J. G., and Bolley, D. S., 1950. *Elastomer Reinforced with a Modified Clay*. National Lead Co. U.S. Patent 2531396.

Chang, Y. W., Kim, S., Kang, S. C., and Bae, S. Y., 2011. Thermomechanical properties of ethylene-propylene-diene terpolymer/organoclay nanocomposites and foam processing in supercritical carbon dioxide. *Korean Journal of Chemical Engineering* 28: 1779–84.

Chang, Y. W., Lee, D., and Bae, S. Y., 2006. Preparation of polyethylene–octene elastomer/clay nanocomposite and microcellular foam processed in supercritical carbon dioxide. *Polymer International* 55: 184–9.

Choudalakis, G. and Gotsis, A. D., 2009. Permeability of polymer/clay nanocomposites: A review. *European Polymer Journal* 45: 967–84.

Chuayjuljit, S., Maungchareon, A., and Saravari, O., 2010. Preparation and properties of palm-oil based rigid polyurethane nanocomposite foams. *Journal of Reinforced Plastics and Composites* 29: 218–25.

Di, Y., Iannace, S., Di Maio, E., and Nicolais, L., 2005. Poly(lactic acid)/organoclay nanocomposites: Thermal, rheological properties and foam processing. *Journal of Polymer Science: Part B: Polymer Physics* 43: 689–98.

Ema, Y., Ikeya, M., and Okamoto, M., 2006. Foam processing and cellular strucuture of poly-lactide-based nanocomposites. *Polymer* 47: 5350–59.

Esteves, A. C. C., Barros-Timmons, A., and Trindade, T., 2004. Nanocompósitos de matriz. polimérica: Estratégias de síntese de matérias híbridos. *Química Nova* 27: 798–806.

Fu, J. and Naguib, H. E., 2006. Effect of nanoclay on the mechanical properties of PMMA/clay nanocomposite foams. *Journal of Cellular Plastics* 42: 325–42.

Guo, Z., Lee, J., and Tomasko, D., 2008. $CO_2$ permeability of polystyrene nanocomposites and nanocomposite foams. *Industrial & Engineering Chemistry Research* 47: 9636–43.

Han, X., Zeng, C., Lee, J., Koelling, K. W., and Tomasko, D. L., 2003. Extrusion of polystyrene nanocomposite foams with supercritical $CO_2$. *Polymer Engineering and Science* 43: 1261–75.

Ibeh, C.C. and Bubacz, M., 2008. Current trends in nanocomposite foams. *Journal of Cellular Plastics* 44: 493–515.

Istrate, O. M. and Chen, B., 2011. Relative modulus-relative density relationships in low density polymer–clay nanocomposite foams. *Soft Matter* 7: 1840–48.

Jiang, X. L., Bao, J. B., Liu, T., Zhao, L., Xu, Z. M., and Yuan, W. K., 2009. Microcellular foaming of polypropylene/clay nanocomposites with supercritical carbon dioxide. *Journal of Cellular Plastics* 45: 515–38.

Jo, C. and Naguib, H. E., 2007a. Constitutive modeling of HDPE polymer/clay nanocomposite foams. *Polymer* 48: 3349–60.

Jo, C. and Naguib, H. E., 2007b. Effect of nanoclay and foaming conditions on the mechanical properties of HDPE-clay nanocomposite foams. *Journal of Cellular Plastics* 43:111–21.

Jo, C. and Naguib, H. E., 2007c. Processing, characterization and modeling of polymer/clay nanocomposite foams. *Journal of Physics: Conference Series* 61: 861–8.

Jordan, J. W., 1949. Organophilic bentonites I. Swelling in organic liquids. *Journal of Physical and Colloid Chemistry* 53: 294–306.

Kaneko, M. L. Q. A., Torriani, I., and Yoshida, I. V. P., 2007. Morphological evaluation of silicone/clay slurries by small-angle/wide-angle X-ray scattering. *Journal of Brazilian Chemistry Society* 18: 765–73.

Kawasumi, M., Kohzaki, M., Kojima, Y., Okada, A., and Kamigaito, O., 1989. *Process for Producing Composite Material*. Kabushiki Kaisha Toyota Chuo Kenkyusho. U.S. Patent 4810734.

Ke, Y. C. and Stroeve, P., 2005. *Polymer-Layered Silicate and Silica Nanocomposites*. Amsterdam: Elsevier B.V..

Keramati, M., Ghasemi, I., Karrabi, M., and Azizi, H., 2012. Microcellular foaming of PP/EPDM/organoclay nanocomposites: The effect of the distribution of nanoclay on foam morphology. *Polymer Journal* 44: 433–8.

Khorasani, M. M., Ghaffarian, S. R., Babaie, S., and Mohammadi, N., 2010. Foaming behavior and cellular strucuture of microcellular HDPE nanocomposites prepared by a high temperature process. *Journal of Cellular Plastics* 46: 173–90.

Lagaly, G., 1981. Characterization of clays by organic compounds. *Clay Minerals* 16: 1–21.

Lagaly, G., 1984. Clay-organic interactions. *Philosophical Transactions of the Royal Society A: Mathematical, Physical & Engineering Sciences* 311: 315–32.

Lee, S. Y., Chen, H., and Hanna, M. A., 2008. Preparation and characterization of tapioca starch-poly(lactic acid) nanocomposite foams by melt intercalation based on clay type. *Industrial Crops and Products* 28: 95–106.

Lee, S. Y. and Hann, M. H., 2008. Preparation and characterization of tapioca starch-poly(lactic acid)-Cloisite Na$^+$ nanocomposite foams. *Journal of Applied Polymer Science* 110: 2337–44.

Lee, Y. H., Park, C. B., Wang, K. H., and Lee, M. H., 2005b. HDPE-clay nanocomposite foams blown with supercritical $CO_2$. *Journal of Cellular Plastics* 41: 487–502.

Lee, L. J., Zeng, C., Cao, X., Han, X., Shen, J., and Xu, G., 2005a. Polymer nanocomposite foams. *Composite Science and Technology* 65: 2344–63.

Liao, X., Zhang, H., and He, T., 2012. Preparation of porous biodegradable polymer and its nanocomposites by supercritical $CO_2$ foaming for tissue engineering. *Journal of Nanomaterials* 2012: 1–12.

Liu, H., Han, C., and Dong, L., 2009. Study of the biodegradable poly($\varepsilon$-caprolactone)/clay nanocomposite foams. *Journal of Applied Polymer Science* 115: 3120–9.

Liu, H., Nakagawa, K., Chaudhary, D., Asakuma, Y., and Tadé, M.O., 2011. Freeze-dried macroporous foam prepared from chitosan/xanthan gum/montmorillonite nanocomposites. *Chemical Engineering Research and Design* 89: 2356–64.

Ma, P., Wang, X., Liu, B., Li, Y., Chen, S., Zhang, Y., and Xu, G., 2012. Preparation and foaming extrusion. *Journal of Cellular Plastics* 48:191–205.

Martini, J. E., 1981. The production and analysis of microcellular foam. Ph.D dissertation. Boston, MA: Massachusetts Institute of Technology.

Mitsunaga, M., Ito, Y., Ray, S. S., Okamoto, M., and Hironaka, K., 2003. Intercalated polycarbonate/clay nanocomposites: Nanostructure control and foam processing. *Macromolecular Materials and Engineering* 288: 543–8.

Mittal, V., 2009. Polymer layered silicate nanocomposites: A Review. *Materials* 2:992–1057.

Modesti, M., Lorenzetti, A., and Besco, S., 2007. Influence of nanofillers on thermal insulating properties of polyurethane nanocomposite foams. *Polymer Engineering and Science* 47: 1351–58.

Moghbeli, M. R. and Shahabi, M., 2011. Morphology and mechanical properties of an elastomeric poly(HIPE) nanocomposite foam prepared via an emulsion template. *Iranian Polymer Journal* 20: 343–55.

NGO, T. T. V., Duchet-Rumeau, J., Whittaker, A. K., and Gerard, J. F., 2010. Processing of nanocomposite foams in supercritical carbon dioxide. Part I: Effect of surfactant. *Polymer* 51: 3436–44.

Okada, A., Fukushina, Y., Kawasumi, M., Inagaki, S., Usuki, A., and Sugiyama, S., 1988. *Composite Material and Process for Manufacturing Same*. Kabushiki Kaisha Toyota Chou Kenkyusho. U.S. Patent 4739007.

Pan, L. Y., Zhan, M. S., and Wang, K., 2011. High-temperature-resistant polyimide/montmorillonite foams by solid blending. *Polymer Engineering and Science* 51: 1397–1403.

Papin, R., 1993. Bentonite. In *Kirk-Othmer Encyclopedia of Chemical Technology*, vol. 3, 339–360. New York: John Wiley & Sons.

Pavlidou, S. and Papaspyrides, C. D., 2008. A review on polymer-layered silicate nanocomposites. *Progress in Polymer Science* 33: 1119–98.

Ray, S. S. and Okamoto, M., 2003a. New polylactide/layered silicate nanocomposites, 6a melt rheology and foam processing. *Macromolecular Materials and Engineering* 288: 936–44.

Ray, S. S. and Okamoto, M., 2003b. Polymer/layered silicate nanocomposites: A review from preparation to processing. *Progress in Polymer Science* 28: 1539–1641.

Schmidt, D., Shah, D., and Giannelis, E. P., 2002. New advances in polymer/layered silicate nanocomposites. *Current Opinion in Solid State and Materials Science* 6: 205–12.

Shen, Y. H., 2001. Preparation of organobentonite using nonionic surfactants. *Chemosphere* 44: 989–95.

Shen, J., Cao, X., and Lee, L. J., 2006. Synthesis and foaming of water expandable polystyrene-clay nanocomposites. *Polymer* 47: 6303–10.

Souza, M. A., Pessan, L. A., and Rodolfo Jr., A., 2006. Nanocompósitos de poli(cloreto de vinila) (PVC)/argilas organofílicas. *Polímeros: Ciência e Tecnologia* 16: 257–62.

Strauss, W. and D´Souza, N. A., 2004. Supercritical $CO_2$ processed polystyrene nanocomposite foams. *Journal of Cellular Plastics* 40: 229–41.

Taki, K., Yanagimoto, T., Funami, E., Okamoto, M., and Ohshima, M., 2004. Visual observation of $CO_2$ foaming of polypropylene-clay nanocomposites. *Polymer Engineering and Science* 44: 1004–11.

Tomasko, D. L., Han, X., Liu, D., and Gao, W., 2003. Supercritical fluid applications in polymer nanocomposites. *Current Opinion in Solid State and Materials* 7: 407–12.

Urbanczyk, L., Alexandre, M., Detrembleur, C., Jérôme, C., and Calberg, C., 2010b. Extrusion foaming of poly(styrene-co-acrylonitrile)/clay nanocomposites using supercritical $CO_2$. *Macromolecular Materials and Engineering* 295: 915–22.

Urbanczyk, L., Calberg, C., Detrembleur, C., Jérôme, C., and Alexandre, M., 2010a. Batch foaming of SAN/clay nanocomposites with $scCO_2$: A very tunable way of controlling the cellular morphology. *Polymer* 51: 3520–31.

Wee, D., Seong, D. G., and Youn, J. R., 2004. Processing of microcellular nanocomposite foams by using a supercritical fluid. *Fibers and Polymers* 5: 160–9.

Weickmann, H., Gurr, M., Meincke, O., Thomann, R., and Mülhaupt, R., 2010. A versatile solvent-free "one-pot" route to polymer nanocomposites and the *in situ* formation of calcium phosphate/layered silicate hybrid nanoparticles. *Advanced Functional Materials* 20: 1778–86.

Yeh, J. M. and Chang, K. C., 2008. Polymer/layered silicate nanocomposite anticorrosive coatings. *Journal of Industrial and Engineering Chemistry* 14: 275–91.

Yeh, J. M., Chang, K. C., Peng, C. W., Lai, M. C., Hung, C. B., Hsu, S. C., Hwang, S. S., and Lin, H. R., 2009. Effect of dispersion capability of organoclay on cellular structure and physical properties of PMMA/clay nanocomposite foams. *Materials Chemistry and Physics* 115: 744–50.

Zandi, F., Rezaei, M., and Kasiri, A., 2011. Effect of nanoclay on the physical-mechanical and thermal properties and microstructure of extruded noncross-linked LDPE nanocomposite foams. *Key Engineering Materials* 471–472: 751–6.

Zeng, C., Han, X., Lee, L. J., Koelling, K. W., and Tomasko, D. L., 2003. Polymer–clay nanocomposite foams prepared using carbon dioxide. *Advanced Materials* 15: 1743–7.

Zhu, B., Zha, W., Yang, J., Zhang, C., and Lee, L. J., 2010. Layered-silicate based polystyrene nanocomposite microcellular foam using supercritical carbon dioxide as blowing agent. *Polymer* 51: 2177–84.

# 8 Carbon Nanotube-Polymer Nanocomposite Aerogels and Related Materials: Fabrication and Properties

*Petar Dimitrov Petrov*

## CONTENTS

## 8.1 INTRODUCTION

Polymer nanocomposites are an important class of materials, which are an alternative to the conventionally filled polymers. These are materials in which nanosized inorganic fillers, typically 1–100 nm in at least one dimension, are individually dispersed in a polymer matrix. Because of the nanometer sizes of the filler, nanocomposites exhibit markedly improved properties as compared to the pure polymers or their conventional composites. These include increased modulus and strength, outstanding barrier properties, improved solvent and heat resistance, and decreased flammability (Lagashetty and Venkataraman, 2005).

Since their discovery in the early 1990s, both multi-walled (MWNTs) and single-walled carbon nanotubes (SWNTs) have received considerable attention due to their high strength and flexibility, large aspect ratio, low mass density, and extraordinary electrical, optical, and thermal properties (Fischer, 2006). Therefore, carbon nanotubes (CNTs) have been considered for a wide range of potential applications, especially as fillers in polymer composites (Ajayan et al., 1994; Moniruzzaman and Winey, 2006; Spitalsky et al., 2010). In order to maximize the advantage of CNTs as effective reinforcement for high-strength polymer nanocomposites, CNTs should not form aggregates, and must be well dispersed to enhance the interfacial interaction with the matrix. Together with the reinforcing effect, uniformly dispersed CNTs can imply a high electrical conductivity of the composite by formation of an interconnected nanotube network inside the polymer matrix at a relatively low content (Ramasubramaniama, Chen, and Liu, 2003). The most common methods for fabricating bulk CNT-polymer nanocomposites are based on mixing CNTs and polymers using solution or melt processing or an *in situ* polymerization approach. In order to enhance nanotube affinity to engineering polymer matrices, chemical modification of CNTs is necessary (Tasis et al., 2006).

Weight reduction of nanocomposites is important in many applications because it results in saving materials and energy, as well as easier manipulation. Formation of pores during synthesis or postprocessing is usually an effective way to reduce both the mass and density of bulk materials. Aerogels are an intriguing kind of foam-like, solid-state material with unique properties such as extremely low density, large open pores, and high surface areas (Pierre, 2011). Aerogels consist of a three-dimensional solid network with a large number of air-filled pores that take up most of the volume. A typical procedure for preparation of aerogels involves transformation of precursor molecules in a liquid into wet gel and subsequent drying using specific techniques (i.e., supercritical drying, freeze drying, etc.,) to avoid substantial shrinkage of the solid network formed and to preserve the porosity of material (Biener et al., 2011). Although most aerogels are fabricated from silica and pyrolized organic polymers, there is an increasing interest in preparation of aerogel composites (Ramakrishnan et al., 2006). In particular, assembling CNTs into aerogels creates a new nanocomposite material that integrates the intriguing properties of CNTs with the unique structure and related properties of aerogels.

In this chapter, the recent advances in the field of carbon nanotube/polymer nanocomposite aerogels and related materials are described. An emphasis is paid to the relationship between the preparation method and the most characteristic properties of these materials such as density, surface area, electrical conductivity, mechanical strength, and so forth.

## 8.2 NANOCOMPOSITE AEROGELS CONTAINING A HIGH CONCENTRATION OF CARBON NANOTUBES

The outstanding properties of CNTs in combination with their light weight have prompted studies oriented toward the fabrication of low-density CNT-based foams (Cao et al., 2005). Unfortunately, very often the realization of the intrinsic nanotube properties in macroscopic forms such as foams has been limited. Free-standing CNT aerogels fabricated from a wet gel based on CNTs dispersed by a low molecular mass surfactant apparently suffer from low mechanical strength, resulting from the weak van der Waals forces responsible for the structural integrity (Bryning et al., 2007). The use of polymer binders to structurally reinforce CNT aerogels typically leads

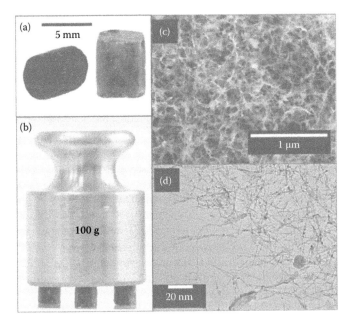

**FIGURE 8.1 (See color insert.)** Images of aerogels. (a) Macroscopic pieces of 7.5 mg/mL CNT aerogels. Pristine CNT aerogel (left) appears black, whereas the aerogel reinforced in a 1 mass% PVA bath (right) is slightly gray. (b) Three PVA-reinforced aerogel pillars (total mass = 13.0 mg) supporting 100 g, or ca. 8000 times their weight. (c) This scanning electron microscopy (SEM) image of a critical-point-dried aerogel reinforced in a 0.5 mass% PVA solution (CNT content = 10 mg/mL) reveals an open, porous structure. (d) This high-magnification transmission electron microscopy (TEM) image of an unreinforced aerogel reveals small-diameter CNTs arranged in a classic filamentous network. (Reprinted from Bryning M. B. et al., 2007, *Adv. Mater.* 19, 661–664, with permission from John Wiley & Sons.)

to a significant increase of mechanical strength. In this case, the amount of polymer is comparable to that of CNTs. For instance, poly(vinyl alcohol) (PVA) was successfully exploited in a number of works for reinforcement of aerogels made from both SWNTs and MWNTs. Bryning and coworkers (2007) reported the first study on the preparation of single- and double-walled CNT/PVA aerogels from aqueous gel precursors by supercritical drying or freeze drying (Figure 8.1).

The nanocomposite aerogels obtained are highly porous materials, with the pore size ranging from tens of nanometers to one micron. Furthermore, the freeze-dried samples possess a second generation of pores from tens to hundreds of micrometers as a result of ice crystals that are formed during the freezing process. It was demonstrated that the addition of a certain amount of PVA (PVA:CNTs mass ratio 1–6) improves the strength and stability of the aerogels as compared to the fragile unreinforced CNT aerogels. Such materials can support at least 8000 times their own mass. The electrical conductivity of the aerogels was found to depend mainly on the drying process and PVA content. The freeze-dried samples are less conductive than the samples dried at supercritical conditions. The difference is attributed to the disruption of the nanotube network that occurs during freezing. The electrical conductivity decreases proportionally to the increase of PVA content in the aerogel and reaches values typical for solid polymer composites with comparable nanotube volume fraction. Interestingly, the same authors discovered that short, high current pulses applied to CNT/PVA nanocomposite aerogel produces an irreversible, stepwise increase of electrical conductivity. The net increase could be quite large, for example, several orders of magnitude, as shown in Figure 8.2.

Although this procedure did not induce any noticeable macroscopic and microscopic changes in the sample, it was suggested that the current pulses decompose PVA at the junction between the nanotubes.

MWNT/PVA nanocomposite aerogels with various nanotube content (25–91 mass%) were fabricated by applying flash freezing of aqueous nanotube/PVA dispersion in liquid nitrogen and subsequent lyophilization (Skaltsas, Avgouropoulos, and Tasis, 2011). During the freezing, ice crystals are formed and, as a result, the dispersed CNTs and PVA macromolecules are accumulated in the regions between crystal galleries. Thus, after sublimation of ice by freeze drying, a CNT/PVA scaffold is produced, whose structure is a negative replica of the ice crystal structure. It is confirmed by SEM analysis and nitrogen adsorption–desorption test that two levels of porosity exist in the aerogels—mesopores identified as the distance between isolated tubes or bundles and macropores identified as the distance between the aerogel walls. A comparative study of aerogels with different MWNT concentration revealed that both the porosity and size of mesopores decrease gradually with decreasing nanotube content.

It was demonstrated that MWNT/PVA aerogels exhibit interesting thermal and catalytic properties as well as absorption capacity of polycyclic aromatic substances and can be applied as catalyst supports, membranes, and thermal conductors.

MWNT/PVA aerogels have been fabricated employing the procedure for synthesis of chemically cross-linked cryogels (Kueseng et al., 2010). In this case, glutaraldehyde was added to CNTs/PVA mixture to cross-link the polymer chains.

Another original approach for fabrication of nanocomposite aerogels based on CNTs and chemically cross-linked polymers was described by Zou, Liu, and Karakoti (2010). The authors first disperse MWNTs by poly(3-hexylthiophene)-b-poly(3-(trimethoxysilyl)propyl methacrylate) (P3HT-PTMSPMA) due to the $\pi$–$\pi$ interaction between nanotubes and P3HT, form a physical gel and, then, cross-link the polymer layer by hydrolysis and condensation of PTMSPMA blocks (Figure 8.3).

It is pointed out that the chemically cross-linked wet gel is a more advantageous precursor for aerogels than the physical gel, which is mechanically weaker, and some physical interactions may become diminished after the removal of the solvent. Further, the wet gel was unidirectionally frozen by liquid nitrogen. Under such conditions, pseudo-steady-state continuous growth of ice crystals generates an array of micrometer-sized polygonal ice rods parallel with the freezing direction. The ice rods act as a template for the honeycomb structure of nanocomposite aerogels obtained by freeze drying. In addition to the unique macroporous structure, it was found that aerogels have mesopores developed in the honeycomb walls, confirming that aerogels are hierarchically porous materials. Noteworthy, the better dispersion of MWNTs in aerogels is indicated by the much narrower mesopore size distribution than that of as-produced MWNTs. Regardless of their ultra-light density (4 mg/cm$^3$), MWNT/

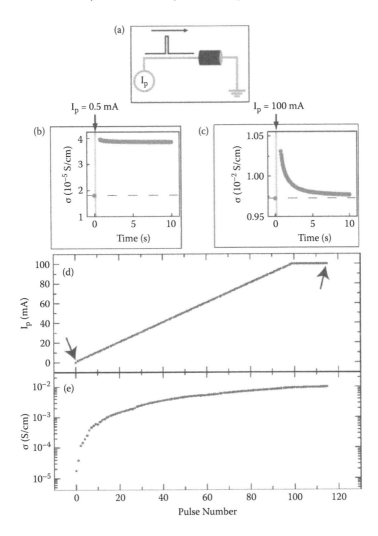

**FIGURE 8.2** (a) Discrete current pulses, $I_p$, applied across a sample, as shown in this schematic, improved the electrical conductivity of PVA-reinforced CNT aerogels. (b) A 15 ms, 0.5 mA current pulse (depicted by the vertical gray line) was applied to a 13.3 mg/mL CNT critical-point-dried sample reinforced in a 0.5 mass% PVA bath. The conductivity improved by ca. 2× after the pulse, and remained at the new level indefinitely. (c) A 15 ms, 100 mA current pulse was applied across the same sample, after the sample had undergone repeated current pulses and the starting resistance was much lower. Although the current here was much higher, conductivity improved only transiently, presumably because of heating, and returned to the original value in a matter of seconds. (d) and (e) The full breakdown history of this sample: (d) pulsing current was linearly increased up to 100 mA (pulse duration was always 15 ms; pulses were spaced more than 30 s apart). Each point represents a discrete current pulse. Arrows point to the pulses depicted in (b) and (c). (e) The corresponding steady-state conductivity, measured 30 s after each pulse. Notice that the conductivity improved by several orders of magnitude after repeated pulsing. (Reprinted from Bryning M. B. et al., 2007, *Adv. Mater.* 19, 661–664, with permission from John Wiley & Sons.)

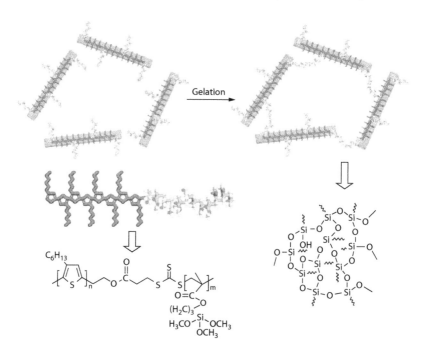

**FIGURE 8.3 (See color insert.)** Schematic illustration of gelation process of P3HT-b-PTMSPMA dispersed MWCNTs. Before gelation, P3HT blocks bond to the MWCNT surface through π–π interaction; PTMSPMA blocks locate at the outer surface of MWCNT. After gelation, MWCNTs interact with each other through chemical bonding formed by PTMSPMA blocks. (Reprinted from Zou J. et al., 2010, *ACS Nano* 4, 7293–7302, with permission from the American Chemical Society.)

P3HT-PTMSPMA aerogels are mechanically robust as confirmed by compression tests with the stress loaded along and perpendicular to the cell axis direction. The material can be repeatedly compressed down to 5% of its original volume and recovers most of its volume after the release of compression. These properties are attributed to the mechanical strength of MWNTs and the cross-linking of polymer binder. The good compression recovery and conductive properties ($3.2\times10^{-2}$ S/cm) of aerogels provided pressure responsive properties of the material (Figure 8.4).

The reproducible and highly sensitive change of the resistance with the applied pressure (5–200 Pa) makes these aerogels a promising candidate for ultrasensitive pressure sensing. In addition, it was found that aerogel's resistance increases when exposed to chloroform vapor in 0.5 s and then completely recovers the previous value in air at a very low detection limit (1 ppb). Thus, such materials have a potential application as chemoresistance vapor sensors.

Stable aerogels based on single- or double-walled CNTs and ferrocene-grafted poly(p-phenyleneethynylene) (Fc-PPE) copolymers were produced by forming a three-dimensional assembly in solution, followed by $CO_2$ supercritical drying (Kohlmeyer et al., 2011). The method is based on the fact that at sufficient copolymer concentration, the ferrocenyl groups act as anchoring units and the

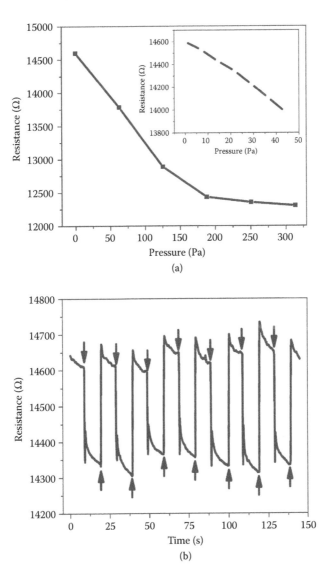

**FIGURE 8.4**   (a) The resistance change of MWCNT aerogel with applied pressure. At low-pressure range, the resistance varies linearly with the pressure (inset). (b) Change of WCNT aerogel resistance with loadings. Arrows indicate the moments of the application and release of loading. (Reprinted from Zou J. et al., 2010, *ACS Nano* 4, 7293–7302, with permission from the American Chemical Society.)

copolymer cross-links CNTs, resulting in formation of a stable organic gel precursor for fabrication of aerogels. The mass of Fc-PPE (20, 33, and 50%) has notable effect on the mechanical stability, electrical conductivity, surface area, and porosity of aerogels. Aerogels with 20 mass% Fc-PPE exhibit the highest electrical conductivity, specific surface area, and mesopore volume, however, their mechanical weakness makes them less attractive materials. Conversely, aerogels with 50% Fc-PPE display the best mechanical properties, but have low specific surface area. Hence, aerogels containing 33% Fc-PPE are considered the most suitable materials for further investigation. While the copolymer is crucial to form stable organogels and aerogels, it was found that it could substantially block the micropores (<2 nm) and small mesopores (2–4 nm) and reduce the specific surface area and mesopore volume of aerogel. However, it was demonstrated that by a thermal annealing procedure the copolymer can be decomposed and the originally blocked micropores and small masopores can be reopened. As a result, specific surface areas of annealed aerogels were increased dramatically by 50–240%, while the electrical conductivity was increased by a factor of 6–13 (up to 1.96 S/cm).

Aerogels based on CNTs and natural polymers have also been investigated for numerous applications. For instance, chitosan was used as both a dispersion agent and structure binder to fabricate MWNT aerogels via the so-called ice segregation-induced self-assembly process (Gutiérrez et al., 2007). This approach allows preparation of aerogels built from interconnected MWNT/chitosan sheets arranged in parallel layers (Figure 8.5) by unidirectional freezing of the aqueous dispersion in liquid nitrogen and subsequent freeze drying.

Aerogel composition determines its physical properties. For example, the increase of MWNT content from 66 to 89 mass% increases the electrical conductivity of material from 0.17 to 2.5 S/cm. Aerogel containing 89 mass% MWNTs was further decorated with Pt nanoparticles and its electrocatalytic properties were studied. Due to the combination of high electrical conductivity, high

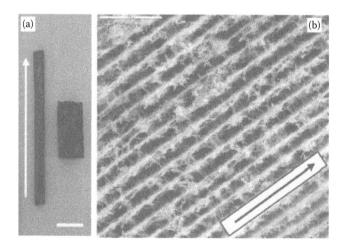

**FIGURE 8.5 (See color insert.)** Picture of MWCNT/CHI monoliths with different shapes and sizes resulting from the ISISA processing of MWCNT/CHI suspensions placed in different disposable containers; an insulin syringe (left) and a polystyrene cuvette (right). (a) Bar is 1 cm. SEM micrograph of the longitudinal section of a MWCNT/CHI monolith. (b) Bar is 50 μm. The MWNT content of every monolith shown is 85 mass%. Arrows indicate the direction of freezing. (Reprinted from Gutiérrez M.C. et al., 2007, *J. Phys. Chem. C* 111, 5557–5560, with permission from the American Chemical Society.)

internal reactive surface, and the easy access through the aligned microchanelled structure, such material has remarkable performance as anode for a direct methanol fuel cell.

The advantages of aerogels with aligned structure over those with disordered structure were demonstrated by Kwon, Kim, and Jin (2009). Aerogels based on MWNTs and silk fibroin with aligned structure were obtained by unidirectional freezing of aqueous dispersions in liquid nitrogen at a constant rate of 10 cm/h and freeze drying, while the nonaligned aerogels were prepared by formation of MWNTs/silk fibroin hydrogels, flash freezing, and freeze drying. As a result, all aligned aerogels have a porous microchannel structure, while the nonaligned aerogels possess a 3D random network structure. Inner morphology observations by TEM revealed that MWNTs in the aligned samples are better connected to each other than those in the disordered samples. Overall, the aerogels with aligned porous structure exhibit higher electrical conductivity, thermal resistance, and surface area as compared to nonaligned aerogels.

Foam-like MWNT/polymer nanocomposite has also been prepared by a method already described for silica foams (Carn et al., 2004). First, SWNTs were dispersed in water using carboxymethyl cellulose sodium salt and then a liquid foam was obtained via a bubbling process on a column (Leroy et al., 2007). This method allows control of the size and shape of cells by playing with the diameter of the porous filter and the liquid flux. Finally, the liquid foam was cooled down to –80°C and lyophilized.

## 8.3   NANOCOMPOSITE AEROGELS CONTAINING A LOW CONCENTRATION OF CARBON NANOTUBES

The general trend seen in the studies focused on nanocomposite aerogels containing a high concentration of CNTs is that the polymers serve as structure binders, that is, they reinforce the material, but do not dominate the physico-mechanical properties. As a typical example, the addition of polymers, which are usually insulators, decreases the electrical conductivity of CNT aerogel to a certain extent but the values reported are still in the range of those accepted for conducting materials.

In terms of conventional CNT/polymer nanocomposites, usually the incorporation of CNT fillers aims at increasing both the mechanical properties and conductivity of polymer matrix. Preferably, the content of CNTs should be as low as possible, thus, providing cheaper materials. In particular, the fabrication of CNT/polymer nanocomposite aerogels with low CNT content (≤5 mass%), which combine good mechanical properties and high electrical conductivity, still seems to be a challenge. One of the first papers dealing with CNT/polymer aerogels highlighted that the resistivity of such material can be decreased significantly only by thermal decomposition of the polymer (Nabeta and Sano, 2005). Otherwise, the aerogels obtained from SWNT/gelatin hydrogels (ca. 1, 3, and 16 mass% nanotubes with respect to polymer) by freeze drying are insulators. According to SEM analysis and electrical resistivity measurements, SWNTs are completely embedded in the polymer matrix but do not form an interconnected network (percolation threshold) even at the highest nanotube concentration studied.

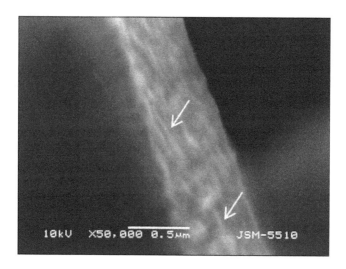

**FIGURE 8.6** Representative SEM micrograph of PNIPAAm cryogel with MWNTs embedded into the walls. (Reprinted from Petrov P. and Georgiev G., 2012, *Eur. Polym. J.* 48, 1366–1373, with permission from Elsevier.)

A similar tendency was observed by our team in experiments focused on the preparation of MWNT/polymer nanocomposites employing an established procedure for synthesis of cryogels via UV irradiation (Petrov and Georgiev, 2012). When MWNTs were mixed with the reagents and then the cryogel was formed by cryogenic treatment and irradiation with UV light, all nanotubes were fully embedded into the cross-linked polymer matrix due to the cryo-structuring phenomena (Figure 8.6).

As expected, at low CNT concentration (0.1–0.5 mass%), the nanocomposite aerogels were insulators, determined by the nature of polymer matrix. One should mention that these materials were fabricated mainly for comparison with other nanocomposite aerogels that exhibit high electrical conductivity at the same CNT's concentration. The last materials were obtained by deposition of CNTs onto the inner surface of preformed aerogels (Petrov and Georgiev, 2011). The process involves soaking of chemically cross-linked polymer aerogels in stable aqueous dispersion of modified CNTs, freezing at –20°C, and subsequent lyophilization. It is assumed that during the cryogenic treatment, most of the water in the system forms ice, whereas CNTs are accumulated on the top of crystals and are gradually pushed to the cryogel walls by the growing crystals (Figure 8.7).

As a result, aerogel walls are coated with a very thin layer of interconnected tubes (Figure 8.8) unlike the aerogels where the tubes are embedded into the polymer matrix.

The layer is built mainly by individual CNTs instead of bundles, which favors the formation of an interconnected CNT network at low concentration. Indeed, electrical conductivity measurements revealed a sharp decrease of aerogel resistivity at CNT concentration in the range of 0.12–0.15 mass%. Since the polymer matrix is made by insulating polymers, the high conductivity values (Table 8.1) are attributed

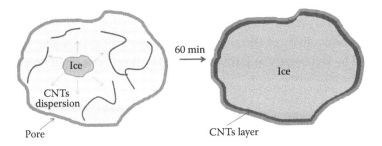

**FIGURE 8.7 (See color insert.)** Schematic representation of ice-mediated deposition of CNTs onto inner surfaces of cryogel walls. (Reprinted from Petrov P. and Georgiev G., 2011, *Chem. Commun.* 47, 5768–5770, with permission of The Royal Society of Chemistry.)

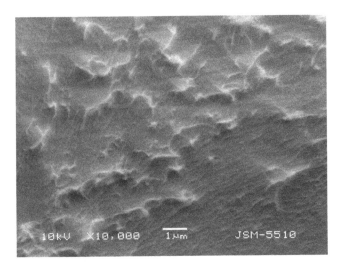

**FIGURE 8.8** Representative SEM micrograph of the wall of a freeze-fractured macroporous PNIPAAm/MWNT composite containing 0.15 mass% CNTs to the polymer weight. (Reprinted from Petrov P. and Georgiev G., 2011, *Chem. Commun.* 47, 5768–5770, with permission of The Royal Society of Chemistry.)

to the CNT's arrangement into a 3D network. It is important to note that nanocomposite aerogels with high electrical conductivity were obtained from different polymer precursors at a similar concentration of both SWNTs and MWNTs. The tensile properties of nanocomposites obtained by deposition of CNTs onto preformed aerogels (Table 8.2) were nearly the same as compared to the native polymer matrix (without CNTs). On the other hand, by changing the polymer type nanocomposite, aerogels of different stiffness and flexibility can be fabricated. For instance, the materials based on poly(N-isopropylacrylamide) (PNIPAAm) and polyacrylamide (PAAm) are harder and more brittle, while those based on 2-hydroxyethylcellulose (HEC) and poly(ethylene oxide) (PEO) are softer and more flexible. Moreover, mixing of a soft (HEC) and a rigid (PAAm) precursor provides material that combines both good stiffness and high flexibility (Figure 8.9). Concerning the conductivity

**TABLE 8.1**

**Electrical Conductivity of Different CNT/Polymer Nanocomposite Aerogels Fabricated by Deposition of Carbon Nanotubes onto Preformed Polymer Aerogels**

| Polymer Type* | CNT Type | CNT Concentration (wt%) | Electrical Conductivity (S/m) |
|---|---|---|---|
| HEC | SWNTs | 0.14 | $5.9 \times 10^{-2}$ |
| HEC/PAAm | SWNTs | 0.12 | $3.2 \times 10^{-2}$ |
| PNIPAAm | SWNTs | 0.15 | $2.8 \times 10^{-2}$ |
| PNIPAAm | MWNTs | 0.15 | $1.1 \times 10^{-2}$ |
| PEO | MWNTs | 0.12 | $5.1 \times 10^{-2}$ |
| PEO/PAAm | MWNTs | 0.15 | $4.2 \times 10^{-2}$ |

* Electrical conductivity of all polymer aerogel precursors was below the instrument detection limit (0.0001 S/m).

**TABLE 8.2**

**Mechanical Properties of Different Precursors and CNT/Polymer Nanocomposite Aerogels Fabricated by Deposition of Carbon Nanotubes onto Preformed Polymer Aerogels**

| Polymer Type | CNT Type | CNT Concentration (wt%) | Yield Strength (MPa) | Young's Modulus (MPa) | Elongation at Ultimate Strength (%) |
|---|---|---|---|---|---|
| PNIPAAm | — | — | $0.021 \pm 0.002$ | $3.2 \pm 0.1$ | $3.4 \pm 0.1$ |
| PNIPAAm | MWNTs | 0.15 | $0.020 \pm 0.001$ | $3.1 \pm 0.1$ | $3.0 \pm 0.1$ |
| HEC | — | — | $0.004 \pm 0.001$ | $0.06 \pm 0.01$ | $21.9 \pm 1$ |
| HEC | SWNTs | 0.14 | $0.004 \pm 0.001$ | $0.06 \pm 0.01$ | $19.8 \pm 1$ |
| HEC/PAAm | — | — | $0.011 \pm 0.002$ | $1 \pm 0.1$ | $10.1 \pm 1$ |
| HEC/PAAm | SWNTs | 0.12 | $0.012 \pm 0.001$ | $1.1 \pm 0.1$ | $9.4 \pm 1$ |

level, such nanocomposite aerogels are suitable for applications as electrostatic dissipation materials (Ramasubramaniam, Chen, and Liu, 2003).

Another strategy for preparation of highly conducting CNT/polymer aerogels with low concentration of CNTs is based on the coupling of CNTs and intrinsically conducting polymers (Zhang et al., 2011). Accordingly, composite aerogels containing MWNTs (1–20 mass%) were obtained by supercritical $CO_2$ drying of poly(3,4-ethylenedioxythiophene)-polystyrenesulfonate (PEDOT-PSS)/MWNTs hydrogel precursors. For instance, the embedding of 5 mass% MWNTs increased the conductivity of PEDOT-PSS aerogel from $1.2 \times 10^{-2}$ to $6.9 \times 10^{-2}$ S/cm. Importantly, it was established that the specific surface area of aerogels increases with CNT content up to 2 mass% and then drops down (Figure 8.10). This result is ascribed to the quality of MWNTs dispersal in the polymer matrix. When loading an amount of

**FIGURE 8.9**   Flexible MWNT/HEC-PAAm nanocomposite aerogel prepared by deposition of nanotubes on to the preformed polymer gel.

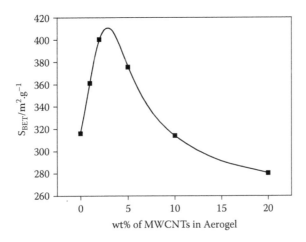

**FIGURE 8.10**   BET surface areas versus nanotube contents of PEDOT–PSS/carbon nanotubes composite aerogels. (Reprinted from Zhang X. et al., 2011, *Carbon* 49, 1884–1893, with permission from Elsevier.)

MWNTs ≤ 2 mass%, nanotubes are dispersed uniformly in the aerogel, so the higher the nanotube loading, the more the individual nanotubes take place in the surface area enhancement. Above 2 mass%, the increase in MWNT concentration leads to worse dispersing and, consequently, the number of individual tubes decreases.

## 8.4   CARBON NANOTUBE-BASED AEROGELS IMPREGNATED WITH POLYMERS

The benefit of fabricating robust, highly conducting CNT-based aerogels possessing open-cell macroporous structure was further exploited toward development of new types of composite materials. They were prepared by backfilling or infiltrating a CNT-based sponge with melted thermoplastic polymer or other liquid (resin, monomer) that could be subsequently cured to form a solid composite comprising a 3D nanotube network embedded in the polymer matrix. Thus, a CNT aerogel (prepared from SWNTs and gelatin and subsequent thermal decomposition of the polymer) was immersed in a low-density polyethylene melt and then cooled (Nabeta and Sano, 2005). The porous structure allowed the polymer to penetrate into the sponge without disintegrating the foam structure. The resulting composite has similar electrical conductivity to the original nanotube aerogel but is more elastic. One of the major differences between the present composite and the composites made by dispersing CNTs in polymers is that the nanotube network is formed prior to mixing and its integrity is not disrupted during the process.

In a similar manner, PVA-reinforced CNT aerogel was successfully impregnated with epoxy resin and cured (Bryning et al., 2007). The initial results indicated that, upon filling and curing, the conductivity of composite remained constant to within a factor of two. Further, this concept was extended to the preparation of stretchable conductors by backfilling SWNT aerogels with the elastomer polydimethylsiloxane (PDMS) (Kim, Vural, and Islam, 2011). The resultant composites are highly bendable, deformable, and stretchable. The electrical conductivity varies in the range of 70–108 S/m and remained the same under high bending strain and increased slightly at a tensile strength of 100%. The transparency of the elastic conductors increased with decrease of the thickness and, for example, 3-μm thin composite film exhibited transmittance of 93% in the visible light region.

Moreover, measurement conducted after 20 stretch–release cycles, both parallel and perpendicular to the conduction direction, revealed that the conductivity remains almost constant. As mentioned above, one of the striking characteristics of SWNT aerogel/PDMS composite films is the negligible dependence of their electrical conductivity on bending deformations (Figure 8.11).

This behavior was attributed to the isotropic structure of SWNT aerogel and short mean length of the nanotubes, which allows it to withstand multiaxial loading and elongation with negligible change in the electrical properties. Importantly, all composite materials remained robust after repeated stretch–release cycles and bending.

Carbon nanotube/polymer composite fibers have been obtained by infiltration of a liquid monomer into millimeter-long MWNT aerogel fibers with subsequent *in situ* polymerization (Zhang et al., 2009). It was found that the composition can be changed by playing with the soaking time. For instance, the soaking of MWNT aerogel fibers in the monomer (methylmethacrylate, MMA) for 3 h yielded composite containing 15 mass% MWNTs. SEM analysis of this composite showed a homogeneous penetration of MMA into the CNT fiber (Figure 8.12).

With an increase in MMA soaking time to 12 h, more liquid penetrates into the CNT fibers and they become swollen. In this case, the concentration of MWNTs in the final composite was 3 mass%. At such low content, there are some regions with heterogeneous distribution of polymer. The composite fibers exhibited improved

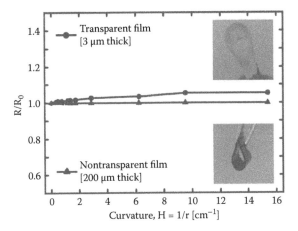

**FIGURE 8.11** Normalized resistance as a function of radius of curvature for transparent and nontransparent SWNT-aerogel/PDMS composite films. For all composite films, the electrical conductivity remained almost unchanged under bending deformations. The insets show transparent and nontransparent SWNT-aerogel/PDMS composite films bent by 180°. (Reprinted from Kim K. H. et al., 2011, *Advanced Materials* 23, 2865–2869, with permission from John Wiley & Sons.)

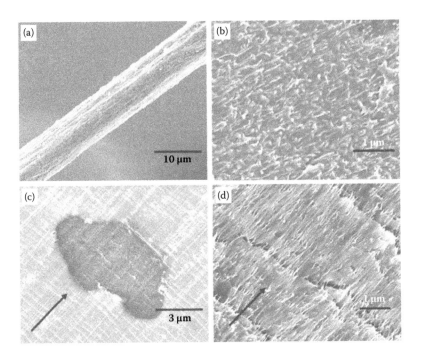

**FIGURE 8.12** SEM images of CNT-PMMA composite fiber with 3 h of MMA liquid soaking: (a) fiber surface, (b) large magnification of (a), (c) cross-section, and (d) large magnification of (c). Arrows show the cutting directions. (Reprinted from Zhang, S. et al., 2009, *Macromol. Rapid Commun.* 30, 1936–1939, with permission from John Wiley & Sons.)

tensile strength and modulus as compared to control PMMA (without CNTs) and as-drowned MWNT aerogel fiber. It was proposed that the local ordering of PMMA (evidenced by WAXS) could increase the load transfer to the aligned 1 mm long CNTs in the composite fibers.

Highly conductive stiff composite with remarkable low CNT concentration (1.2 mass%) were fabricated from SWNT-based aerogel and PDMS (Worsley, Kucheyev, Kuntz et al., 2009). The composite material exhibited ca. 300% increase in the elastic modulus relative to the unloaded PDMS elastomer and electrical conductivity over 1 S/cm. SEM study showed that the SWNT network is homogeneously distributed throughout the polymer matrix, suggesting that there is good wetting at the PDMS/nanotube interface and that the CNT scaffold is intact after infiltration and curing.

## 8.5 AEROGELS FORM CARBON NANOTUBE/ POLYMER NANOCOMPOSITE PRECURSORS

As already mentioned, macroscopic low-density aerogels fabricated from CNTs, though conductive, are mechanically weak and the use of polymer binders to reinforce the foam-like structure is required (Bryning et al., 2007). The binders often decrease the electrical and thermal properties of the CNT scaffold due to their insulating nature. The design of a CNT aerogel with a conductive binder, however, could yield materials, which simultaneously exhibit the desirable properties like mechanical robustness and high electrical conductivity. A valuable strategy for fabricating such composite materials, based on the carbonization of a polymer binder, has been reported by Worsley, Satcher, and Baumann (2008). It is based on the *sol-gel* method typically used for the synthesis of carbon aerogels (Kong et al., 1993). The preparation protocol involves mixing CNTs with resorcinol, formaldehyde, and catalyst ($NaCO_3$) in water and formation of hydrogel due to the polycondensation reaction of resorcinol and formaldehyde. Then, the hydrogel is supercritically dried with $CO_2$. The obtained composite CNT/polymer aerogel precursor was pyrolysed at 1050°C in an inner atmosphere to afford a CNT/carbon nanocomposite aerogel (CNT/CA). An example of double-walled carbon nanotube (DWNTs)/CA containing relatively low amounts of nanotubes is shown in Figure 8.13. The structure of aerogel consists of an interconnected network of primary carbon particles and embedded DWNTs.

It was established that the electrical conductivity of DWNTs/CA increases with nanotube concentration and reaches 8.1 S/cm at the highest content studied (8 mass%). A similar trend was observed by the same authors for the thermal conductivity of DWNTs/CA (Worsley, Satcher, and Baumann, 2009). Further, the study was extended to the fabrication of SWNTs/CA containing up to 55 mass% nanotubes (Worsley, Pauzauskie et al., 2009). Interestingly, above 20 mass%, the microstructure of aerogel changed from a network of carbon nanoparticles to a network of randomly oriented filament-like formations. Hence, at high SWNT concentrations, the majority of nanoparticles are located on the surface of the SWNTs and interconnect the nanotube bundles. In other words, above concentrations of 20 mass%, SWNTs dominate the microstructure of aerogels and determine their

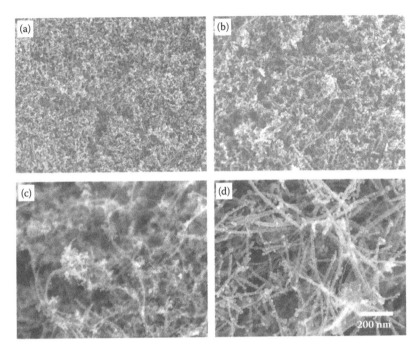

**FIGURE 8.13** SEM images of SWNT-CA with (a) 4 mass%, (b) 20 mass%, (c) 30 mass%, and (d) 55 mass% SWNTs. (Reprinted from Worsley M. A., Pauzauskie P. J. et al., 2009, *Acta Mater.* 57, 5131–5136, with permission from Elsevier.)

properties. For instance, a pronounced enhancement of the electrical conductivity and large decrease of the volumetric shrinkage that occurs during the drying and carbonization steps were observed for SWNT/CA containing 30 and 55 mass% nanotubes as compared to the aerogels with less than 20 mass% nanotubes. Nanoindentation measurements of SWNT/CAs showed density dependences of mechanical properties (Worsley, Kucheyev, and Satcher et al., 2009). On the other hand, the morphology change from nanoparticle network to filament network made by CNT bundles interconnected by carbon particles, resulted in an increase of the Young's modulus by a factor of 2 to 3 (Shin et al., 2012). In addition, these materials exhibited remarkable elastic behavior up to compressive strains as large as 80% (Figure 8.14, inset). Compared to other types of aerogels, at given density, CNT/CAs seems to be the stiffest material. For example, at density less than 100 mg/cm, CNT/CAs with high nanotube content are ca. 12 times stiffer than conventional silica aerogel (Figure 8.14).

The use of MWNTs/polyacrylonitrile as an alternative precursor for preparation of MWNT/CAs has been reported by Bordjiba, Mohamedi, and Dao (2007). The amount of MWNTs in nanocomposite aerogels was varied from 3 to 10 mass%. These materials also exhibited high specific surface area, specific capacitance, and conductivity, and were considered for electrochemical capacitor applications.

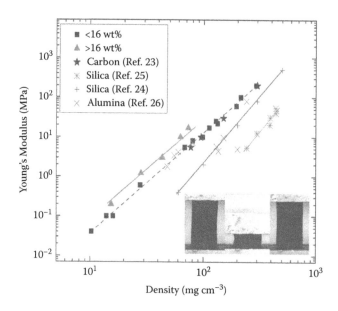

**FIGURE 8.14** Dependence of Young's modulus on density for monolithic CNT foams compared to carbon, silica, and alumina aerogels. The inset shows the sequence of uniaxial compression of a monolith (30 mg/cm³ and 55 mass% CNT content), illustrating the superelastic behavior with complete strain recovery after compression to strains as large as ~76%. (Reprinted from Worsley M. A., Kucheyev S. O. et al., 2009, *Appl. Phys. Lett.* 94, 073115, with permission from the American Institute of Physics.)

## 8.6  CONCLUSIONS

Fabricating CNT-polymer nanocomposite aerogels is a straightforward strategy to obtain ultralight foam-like materials with intriguing properties and many potential applications as vapor sensors, catalyst support, thermal conductors, anodes, electrostatic dissipation materials, and so forth. As a whole, CNT-polymer nanocomposite aerogels possess high surface area and electrical conductivity, very low density, and reasonable mechanical strength. Usually, the polymer fraction in aerogels is comparable to CNT mass and acts as a structural binder. Chemical cross-linking of polymers in the wet gel precursor leads to enhanced mechanical properties of aerogels. Alternatively, nanocomposite aerogels with low CNT content and high conductivity can be prepared by ice-mediated deposition of modified nanotubes onto preformed polymer aerogels. CNT-polymer nanocomposite aerogels can be also used as precursors for preparation of monolith materials, by infiltration with melted polymer, monomer, or resin, or CNT-carbon aerogels by pyrolysis of the polymer fraction.

## REFERENCES

Ajayan, P. M., Stephan, O., Colliex, C., and Trauth, D. 1994. Aligned carbon nanotube arrays formed by cutting a polymer resin-nanotube composite. *Science* 265:1212–1214.

Biener, J., Stadermann, M., Suss, M. et al. 2011. Advanced carbon aerogels for energy applications. *Energy Environ. Sci.* 4: 656–667.

Bordjiba, T., Mohamedi, M., and Dao, L. H. 2007. Synthesis and electrochemical capacitance of binderless nanocomposite electrodes formed by dispersion of carbon nanotubes and carbon aerogels. *J. Power Sources* 172:991–998.

Bryning, M. B., Milkie, D. E., Islam, M. F., Hough, L. A., Kikkawa, J. M., and Yodh, A. G. 2007. Carbon nanotube aerogels. *Adv. Mater.* 19:661–664.

Cao, A., Dickrell, P. L., Sawyer, W. G., Ghasemi-Nejhad, M. N., and Ajayan, P. M. 2005. Materials science: Super-compressible foamlike carbon nanotube films. *Science* 310:1307–1310.

Carn, F., Colin, A., Achard, M. F., Deleuze, H., Saadi, Z., and Backov, R. 2004. Rational design of macrocellular silica scaffolds obtained by a tunable sol-gel foaming process. *Adv. Mater.* 16:140–144.

Fischer, J. E. 2006. Carbon nanotubes: Structure and properties. In *Carbon Nanomaterials*, ed. Y. Gogotsi, 41–75. New York: CRC Press/Taylor & Francis.

Gutiérrez, M. C., Hortigüela, M. J., Manuel Amarilla, J., Jiménez, R., Ferrer, M. L., and Del Monte, F. 2007. Macroporous 3D architectures of self-assembled MWCNT surface decorated with Pt nanoparticles as anodes for a direct methanol fuel cell. *J. Phys. Chem. C* 111:5557–5560.

Kim, K. H., Vural, M., and Islam, M. F. 2011. Single-walled carbon nanotube aerogel-based elastic conductors. *Advanced Materials* 23:2865–2869.

Kohlmeyer, R. R., Lor, M., Deng, J., Liu, H., and Chen, J. 2011. Preparation of stable carbon nanotube aerogels with high electrical conductivity and porosity. *Carbon* 49:2352–2361.

Kong, F. M., LeMay, J. D., Hulsey, S. S., Alviso, C. T., and Pekala, R. W. 1993. Gas permeability of carbon aerogels. *J. Mater. Res.* 8:3100–3105.

Kueseng, P., Thammakhet, C., Thavarungkul, P., and Kanatharana, P. 2010. Multiwalled carbon nanotubes/cryogel composite, a new sorbent for determination of trace polycyclic aromatic hydrocarbons. *Microchem. J.* 96: 317–323.

Kwon, S. M., Kim, H. S., and Jin, H. J. 2009. Multiwalled carbon nanotube cryogels with aligned and non-aligned porous structures. *Polymer* 50:2786–2792.

Lagashetty, A. and Venkataraman, A. 2005. Polymer nanocomposites. *Resonance* 10:49–60.

Leroy, C. M., Carn, F., Backov, R., Trinquecoste, M., and Delhaes, P. 2007. Multiwalled-carbon-nanotube-based carbon foams. *Carbon* 45:2317–2320.

Moniruzzaman, M. and Winey, K. I. 2006. Polymer nanocomposites containing carbon nanotubes. *Macromolecules* 39:5194–5205.

Nabeta, M. and Sano, M. 2005. Nanotube foam prepared by gelatin gel as a template. *Langmuir* 21:1706–1708.

Petrov, P. and Georgiev, G. 2011. Ice-mediated coating of macroporous cryogels by carbon nanotubes: A concept towards electrically conducting nanocomposites. *Chem. Commun.* 47: 5768–5770.

Petrov, P. and Georgiev, G. 2012. Fabrication of super-macroporous nanocomposites by deposition of carbon nanotubes onto polymer cryogels. *Eur. Polym. J.* 48:1366–1373.

Pierre, A. C. 2011. History of aerogels. In *Aerogels Handbook*, eds. M. A. Aegerter, N. Leventis, and M. M. Koebel, 3–18. New York: Springer Science and Business Media.

Ramakrishnan, K., Krishnan, A., Shankar, V., Srivastava, I., Singh, A., and Radha, R. 2006. Modern Aerogels. http://www.dstuns.iitm.ac.in/teaching-and-presentations/teaching/undergraduate%20courses/vy305-molecular-architecture-and-evolution-of-functions/presentations/presentations-2007/seminar-2/P4.pdf. (accessed September 21, 2012).

Ramasubramaniama, R., Chen, J., and Liu, H. 2003. Homogeneous carbon nanotube/polymer composites for electrical applications. *Appl.Phys. Lett.* 83:2928–2930.

Shin, S., Kucheyev, S., Worsley, M., and Hamza, A. 2012. Mechanical deformation of carbon-nanotube-based aerogels. *Carbon* 50:5340–5342.

Skaltsas, T., Avgouropoulos, G., and Tasis, D. 2011. Impact of the fabrication method on the physicochemical properties of carbon nanotube-based aerogels. *Microporous and Mesoporous Materials* 143: 451–457.

Spitalsky, Z., Tasis, D., Papagelis, K., and Galiotis, C. 2010. Carbon nanotube–polymer composites: Chemistry, processing, mechanical and electrical properties. *Prog. Polym. Sci.* 35:357–401.

Tasis, D., Tagmatarchis, N., Bianco, A., and Prato, M. 2006. Chemistry of carbon nanotubes. *Chem. Rev.*106:1105–1136.

Worsley, M. A., Kucheyev, S. O., Kuntz, J. D., Hamza, A. V., Satcher Jr., J. H., and Baumann, T. F. 2009. Stiff and electrically conductive composites of carbon nanotube aerogels and polymers. *J. Mater. Chem.* 19:3370–3372.

Worsley, M. A., Kucheyev, S. O., Satcher, J. H., Hamza, A.V., Baumann, T. F. 2009. Mechanically robust and electrically conductive carbon nanotube foams. *Appl. Phys. Lett.* 94: 073115.

Worsley, M. A., Pauzauskie, P. J., Kucheyev, S. O. et al. 2009. Properties of single-walled carbon nanotube-based aerogels as a function of nanotube loading. *Acta Mater.* 57: 5131–5136.

Worsley, M. A., Satcher Jr., J. H., and Baumann, T. F. 2008. Synthesis and characterization of monolithic carbon aerogel nanocomposites containing double-walled carbon nanotubes. *Langmuir* 24:9763–9766.

Worsley, M. A., Satcher, J. H., and Baumann, T. F. 2009. Enhanced thermal transport in carbon aerogel nanocomposites containing double-walled carbon nanotubes. *J. Appl. Phys.* 105:084316.

Zhang, X., Liu, J., Xu, B., Su, Y., and Luo, Y. 2011. Ultralight conducting polymer/carbon nanotube composite aerogels. *Carbon* 49:1884–1893.

Zhang, S., Zhu, L., Wong, C. P., and Kumar, S. 2009. Polymer-infiltrated aligned carbon nanotube fibers by *in situ* polymerization. *Macromol. Rapid Commun.* 30:1936–1939.

Zou, J., Liu, J., and Karakoti, A. S. 2010. Ultralight multiwalled carbon nanotube aerogel. *ACS Nano* 4:7293–7302.

# 9 Nanocomposite Foams from High-Performance Thermoplastics

*Luigi Sorrentino and Salvatore Iannace*

## CONTENTS

## 9.1 INTRODUCTION

Conventional polymeric foams are employed in many application fields such as packaging, buildings industry, transportation (automotive, aeronautics, naval), and human safety because they are lightweight structures that can combine structural properties (stress mitigation, cushion, energy absorption) to functional ones (thermal or electric insulation, acoustic insulation or absorption, weight reduction, comfort management, filtration, thermal stability, fire resistance). The service temperatures in these applications are not much different from room temperature (Gibson and Ashby, 1997) and foams based on conventional polymers fulfill the performance requirements. In this context, the growing needs of reducing weight at high service temperatures (Sun and Mark, 2002; Sun, Sur, and Mark, 2002) for high-performance applications such as in transportation (sandwich structures to be employed in automotive or aeronautics) or electronics (where electric insulation should be combined to high thermal conduction) is leading to wider requests for high-performance foams.

High-performance thermoplastic polymers are gaining increasing attention for producing foamed materials due to their several advantages on thermosets, such as higher impact strength, recyclability, weldability, absence of volatile organic compounds, reduced processing time, and lower manpower for large scale productions (conversely the thermoplastic process is not affordable in case of small productions or prototyping), but a deep understanding of the foaming process is still ongoing (Sorrentino, Aurilia, and Iannace, 2011). Thermoplastics exhibit a very high viscosity when compared to thermoset polymer precursors and have to be processed above their characteristic transition temperatures (glass transition temperature, for amorphous matrices, or melting temperature, in case of semicrystalline polymers). Unlike thermosets, they can also be worked several times because cross-linking does not take place and molecules can still flow above specific temperatures.

High-performance thermoplastics (HPTP) have a narrow market diffusion due to their high cost and the difficulties in their processing. Consequently, HPTPs are only used in specialized applications that require a polymeric matrix with a combination of extraordinary properties, from both a mechanical point of view (high stiffness, strength, and toughness) and a functional one. In particular, HPTPs are very interesting for their thermal properties (service temperatures, thermal stability, or shrinkage properties) and their FST characteristics, being frequently intrinsically flame retardant.

Among all the peculiar properties, which are conventionally used to distinguish HPTPs from commodity or engineering polymers, the transition temperatures are the most relevant. In fact, the glass transition temperature, $T_g$, and the melting temperature, $T_m$, can reach values as high as 270°C for amorphous polymers and 380°C for semicrystalline polymers, respectively, extending their working temperatures to the lower limit of metals.

The main properties of high-performance thermoplastics are summarized in Table 9.1 and compared to some commodity polymers and common metals. It is evident that from a mechanical point of view, HPTPs exhibit improved properties with respect to conventional polymers, but that they are far from the values of metals even if looking to specific properties. HPTPs, of course, can be shaped and are characterized by processing temperatures much lower than metals.

To further improve HPTP's structural as well as functional properties, additives can be incorporated into the polymeric matrix. Short fibers from glass, aramid, carbon, or basalt are often used to improve the mechanical properties of HPTPs, and to extend their working temperature range due to the increase of the heat deflection temperature. They are usually dispersed in the melt state through extruders or mixers, but due to the high temperature needed, the processing conditions have to be carefully controlled. The elastic moduli increase is proportional to the fiber content, but a high amount of short fibers (25% or more) have to be used to gain significant

## TABLE 9.1
## Main Properties of High-Performance Thermoplastic Polymers

| Material | Heat Deflection Temperature @ 1.82 MPa, °C | Tensile Yield Strength MPa | Tensile Modulus GPa | Impact Strength (Izod Notched) J/m | Density g/cm3 |
|---|---|---|---|---|---|
| HDPE | 44 | 28 | 1.3 | 32 | 0.96 |
| PP | 52.8 | 22 | 1.3 | 30 | 0.91 |
| PC | 129 | 62 | 2.4 | 850 | 1.20 |
| PMMA | 92 | 72.4 | 3.0 | 21 | 1.19 |
| PEK | 163 | 104 | 3.2 | — | 1.27 |
| PEEK | 160 | 94 | 3.7 | 84 | 1.32 |
| PEKK | 175 | 103 | 4.1 | 55 | 1.30 |
| PET | 65 | 80 | 2.6 | 90 | 1.37 |
| PEN | 109 | 108 | 2.5 | 39 | 1.33 |
| PA66 | 90 | 83 | 2.8 | 53 | 1.14 |
| LCP | 187 | 182 | 10.6 | 95 | 1.40 |
| PAEK | 160 | 95 | 3.2 | 64 | 1.38 |
| PCT | 66.2 | 48 | 1.8 | 150 | 1.21 |
| PES | 234 | 84 | 2.4 | 86 | 1.37 |
| PPS | 115 | 86 | 3.2 | 17 | 1.35 |
| PI | 238 | 92 | 2.8 | — | 1.33 |
| PAI | 265 | 138 | 4.0 | 111 | 1.38 |
| PEI | 210 | 105 | 3.0 | 53 | 1.27 |
| sPS | 95 | 38 (yield) | | — | 1.02 |

| | Melt Temperature °C | Tensile Strength MPa | Tensile Modulus GPa | Impact Strength J/m | Density g/cm$^3$ |
|---|---|---|---|---|---|
| Al | 660 | 17 | 68.0 | | 2.69 |
| Steel | 1430 | 250 | 201.0 | | 7.86 |
| Ti | 1650–1670 | 170 | 116.0 | | 4.50 |

improvements of the heat distortion temperature. Inorganic powders are instead used, still in high amount, to improve tribological properties, flame retardancy, thermal conductivity, EM shielding, electrical conductivity, or to improve crystallization kinetics. A high filler loading creates processing difficulties, related to the good dispersion of the reinforcing phase to avoid aggregates and to the preservation of shape factors (in particular for fibers which must have minimum length to diameter ratios to be effective).

In order to overcome this issue, nanofillers are increasingly considered because they can boost performances with a small weight fraction (lower than 5% by weight). They can also be used to embed functional properties, such as flame retardancy, EM shielding, gas barrier, and can help in controlling the processing parameters, such as crystallization kinetics and foamability. In order to exploit their potential, a very good dispersion is essential, and the processing issues typical of nanocomposite polymers still apply to the HPTP, such as tightening the industrial production processes. Furthermore, property improvements strongly depend on the degree of nanofiller dispersion and the nanofiller–matrix interfacial adhesion (Díez-Pascual, Naffakh, Marco, Ellis et al., 2012; Moniruzzaman and Winey, 2006; Thostenson, Ren, and Chou, 2001).

Conventional platelet-like nanofillers, such as montmorillonite or fluorhectorite nanofillers based on organomodified ammonium salts, cannot be used for either cell nucleation or performance improvement at high processing temperatures because degradation of the organic modifier occurs. Several difficulties are encountered in the production of high-performance nanocomposite-based organomodified nanofillers, in particular through melt blending techniques. In fact, due to the very high processing temperatures needed, nonconventional organic modifiers or a different class of nanofillers should be considered for the preparation of high-performance nanocomposites and their foams.

Structural and functional properties are widely dependent on particle characteristics, such as their amount, nature, shape, particle size, surface characteristics, degree of dispersion in the polymer, and, eventually the size of their aggregates. The huge interfacial area in a polymer nanocomposite helps to influence the composite's properties to a great extent even at rather low filler loading but the homogeneous dispersion of nanoparticles is very difficult to achieve because they tend to easily agglomerate.

High-performance foams based on thermoplastics are difficult to produce by means of conventional thermoplastics technologies, such as the extrusion foaming process, because the high temperatures involved could be responsible for thermal degradation of the matrix, and a careful control of the foaming conditions should be assured. High-performance foams could also be formed starting from prefoamed sheets by means of the thermoforming technology, but only simple geometries can actually be produced due to the high temperatures issue.

To further improve performances, nanometric fillers are widely investigated for embedding in the polymeric matrix. Especially in foams, they can improve cell nucleation, crystallization kinetics, and elongational viscosity (fundamental for the production of a fine cellular structure).

Many studies have been conducted on processes and characterizations of foams from high-performance thermoplastic polymers and their nanocomposites in order to exploit their potential, which is mainly related to the cellular morphology. The morphology of foams is very complex to control, because it is characterized by several parameters, namely specific density, mean cell size, cell size distribution, degree of cell interconnections (with the extremes of the range being closed cells and open cells), and cell anisotropy. Foam characteristics, and in particular the morphology, has to be carefully designed according to the requirements of the specific application. For example, foams with microcellular closed-cell structure (cell density of at least $10^9$ cells/cm$^3$ and diameters of less than 10 µm) (Colton and Suh, 1986,1987a; Sorrentino, Di Maio, and Iannace, 2010) are needed to maximize mechanical or thermal insulating properties, while open-cell morphology with a high degree of interconnections has to be used in filtration applications, with mean cell size proportional to the minimum particle diameter to be sieved.

Actually, very few high-performance foams are available on the market, probably due to the difficulties of manufacturing. They are often based on thermoset polymers and usually produced through complex and long processes, such as foaming from low-molecular-weight precursors or stabilizing the cellular structure by polymer cross-linking. Nanocomposite foams from high-performance thermoplastics are their most promising replacements because they can couple high heat deflection temperatures with a fine control of their morphology, in turn maximizing the performances to the specific application.

## 9.2   PROCESSING PARAMETERS

In order to form polymeric foams, bubbles must first nucleate and grow within the molten or plasticized viscoelastic material. Subsequently, consolidation of the structure must occur due to the increase of viscosity during cooling and/or reduction of plasticization and finally solidification of the continuous phase. The initial nucleation is induced by a change in thermodynamic conditions, generally a change of temperature and/or pressure. During this stage, a second phase is generated from a metastable polymer/gas homogeneous mixture (Cotugno et al., 2005; Klempner and Sendjarevic, 2000).

Several industrial processes are available to produce foamed polymers and are based on the following operations:

- Solubilization of the blowing agent.
- Bubble nucleation generally obtained by a sudden modification of the local thermodynamic conditions that induce a phase separation in the matrix–gas system. Usually, this can be achieved by a decrease of pressure or an increase of temperature.
- Bubble growth through the diffusion of gases from the polymer/gas mixture to the nucleated bubbles.
- Stabilization of the cellular structure through a fine control of the solidification process to avoid the collapse of the structure and/or the coalescence of cells.

The process parameters that can be controlled to tailor the architectural morphology of cellular structures (Sorrentino, Di Maio, and Iannace, 2010; Sun and Mark, 2002; Sun, Sur, and Mark, 2002) such as density, cell size, cell size distribution and cell density (number of cells per unit volume) are mainly (i) the amount of the blowing agent in the matrix/gas mixture, (ii) the foaming temperature, (iii) the pressure drop, and (iv) the pressure drop rate. They should be chosen by taking into consideration the intrinsic limitations related to the particular molecular structure, the viscosity of the melt and, for crystalline materials, their crystallization kinetics (Ramesh, Rasmussen, and Campbell, 1991).

In order to improve the number of cells, it is industrial practice to use nucleating agents to have a fine control of the final foam morphology (Fujimoto et al., 2003; Mitsunaga et al., 2003). Traditional nucleating agents are microsized particles, such as talc, which is cheap, efficient, easy to disperse in the polymer, and in large amounts to form master batches. The presence of this filler gives heterogeneous nucleation, which is characterized by both reduction of the barrier energy to nucleation and improvement of the nucleation sites (Lee et al., 2005). In the foaming process, according to classical nucleation theory (Colton and Suh, 1987a,b,c), the heterogeneous nucleation rate is a function of the heterogeneous nucleation sites' concentration (defined as the number of particles per cubic centimeter). Therefore, the particle's concentration is very important because it sets the upper (theoretical) limit for heterogeneous nucleation. In heterogeneous nucleation, the highest nucleation efficiency can only be achieved when the nucleation on the nucleant surface is energetically favored (relative to its homogeneous counterpart) and the nucleant is homogeneously and effectively dispersed in the polymer matrix (Figure 9.1). In most cases, the observed cell density is much lower than the potential nucleant density, implying that either the nucleants are not energetically effective, or their effects have been compromised due to poor dispersion.

Once the cell is nucleated, it will grow due to the diffusion of the gas molecules from the solution to the gas phase. The phenomena governing the expansion of the cell are strictly related to the transport properties of the gas in the polymeric melt and to the rheological properties of the material around the cell. It is thus necessary to know the dependence of thermodynamic, transport, rheological properties, and surface energy of the gas/polymer mixture to solve the predictive model for the cell growth (Arefmanesh, Advani, and Michaelides, 1992). While the main governing parameter for the formation of cells is the degree of the gas supersaturation, when pressure is suddenly reduced or the temperature is raised, the cell growth is kinetically governed by the gas diffusion from the material/gas solution to the gas phase and by the viscoelastic forces around the growing cell. By controlling nucleation

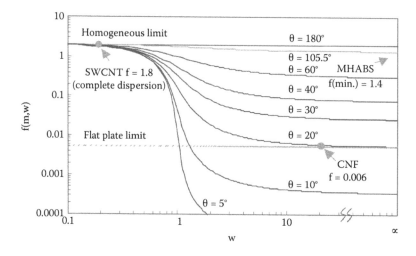

**FIGURE 9.1** Effect on the energy reduction factor f (m, w) of the contact angle θ and the relative curvature w. (Reported from Shen J. et al., 2005, *Polymer* 46, 14, June, 5218–5224, doi:10.1016/j.polymer.2005.04.010, http://linkinghub.elsevier.com/retrieve/pii/S0032386105004349.)

rate and growing rate of cells, it is possible to optimize the morphology of the foam (number and size of cells) and, as a consequence, the mechanical and functional performances of the cellular structure.

Nanometric fillers gained attention due to their potential improvement in different properties when compared to virgin polymer or conventional microcomposites (Alexandre and Dubois, 2000; Giannelis, Krishnamoorti, and Manias, 1999). Due to the high nucleation efficiency, nanoparticles provide a powerful way to increase cell density and reduce cell size. This is particularly beneficial for the production of microcellular foams. Microcellular foams have been considered a lightweight and high strength material for structural applications. However, the narrow operation window and less than desirable cell morphology has limited the applications of this technology. The use of nanoparticles has been investigated to address this difficulty and may greatly enhance the industrial applications of microcellular foam (Di et al., 2005; Di Maio and Iannace, 2009; Lee et al., 2005; Marrazzo, Di Maio, and Iannace, 2008). However, mass production of polymer nanocomposites and nanocomposite foams depends on reliable and affordable synthetic and processing methods. There must be robust techniques to prepare exfoliated nanocomposites with the required mechanical properties in large quantity and low cost. For foam products, various desirable cell morphologies (e.g., small versus large cells, open versus closed cells) must be attainable through the successful control of nucleation and growth of bubbles.

Suitable materials for industrial foaming applications must display adequate properties, and their manufacture process must be relatively simple and inexpensive. The manufacture of foamed products requires a careful combination of polymer/ foaming agent system and the proper optimization of the individual steps of the process. The main requirement is the foamability, which for thermoplastic polymers is related to the rheological characteristics of the melt. Polymers whose viscosity decreases slowly with the increase in temperature are favored and therefore amorphous polymers are generally easier to foam than semicrystalline polymers. In the latter case, materials partially cross-linked and/or highly branched should be used in order to get suitable elongational melt viscosity to withstand the stresses on the cell walls during the growing of the gas bubbles. The knowledge of rheology of polymer/ gas solutions is important for several reasons. It gives important information for the optimization of the manufacturing processes based on extrusion (extrusion foaming, injection molding) or for computer-assisted die design.

The plasticization effect (dependent on the sorption properties of the polymer) of the gas in the polymer melt results in lower viscosity and different rheological behavior, which influences the temperature and pressure profile in the extruder and the die during processing. The extent of viscosity reduction depends upon the polymer/gas pair and the processability of the new foam compound, which can be optimized by adjusting the composition and the processing parameters according to the rheological properties of the mixture (Krause, Mettinkhof et al., 2001; Tang, Huang, and Chen, 2004).

Rheology also plays a fundamental role in the development of the foam morphology. A low viscosity is generally desired for bubble nucleation but strain hardening elongational behavior is necessary for the stabilization of the growing bubbles.

The elongational viscosity, especially its nonlinear response at high strain, is a critical property for the foaming of semicrystalline polymers.

The elongational viscosity is the most important parameter to obtain low density foams. The elongational viscosity can be improved by using long chain branching, polydispersity, and bimodality of the molecular weight distribution, with the presence of high-molecular-weight polymer component. For example, polyesters such as poly(ethylene terephthalate) (PET) (see Figure 9.2a), can be chemically modified with pyromellitic dianhydride (PMDA) (see Figure 9.2b) in order to obtain polymers with a higher molecular weight and branching (Sorrentino et al., 2005). The chain extender can react with the hydroxylic end group of two different PET macromolecules and form two carboxylic groups (see Figure 9.2c). At high temperatures these carboxylic groups can react with -OH groups leading to a branched structure (Bratychak et al., 2002; Chae et al., 2001; Rhein and Ingham, 1973).

However, commercial chain extended PET showed slower crystallization kinetics as compared to that of bottle grade PET. In particular, the energetic barriers to nucleation and molecular mobility of the chain moving from the melt to the growing crystals were higher for the chain extended PET, suggesting a lower nucleation rate and a lower molecular mobility (see Figure 9.3).

The macromolecular structure can be therefore modified to optimize the rheological properties and the crystallization kinetics that are both affected by the degree of cross-linking/branching.

Another potential system to improve the rheological characteristics of HPTP polymers is the use of nanoparticles (Kim and Macosko, 2008). The development

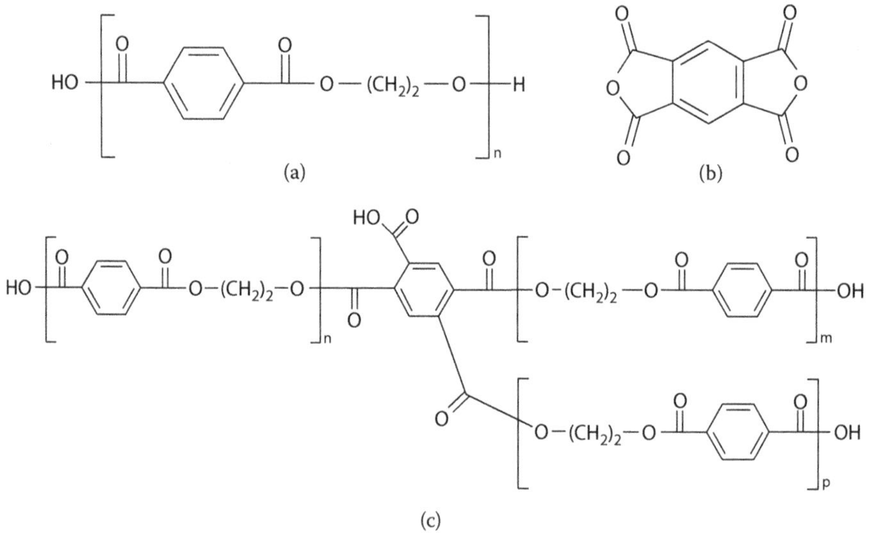

**FIGURE 9.2**   (a) Schematic formulas of PET, (b) PMDA, and (c) branching of PET chains by PMDA. (Reported from Sorrentino L. et al., 2005, *Journal of Polymer Science Part B: Polymer Physics* 43, 15, August 1966–1972, doi:10.1002/polb.20480, http://doi.wiley.com/10.1002/polb.20480.)

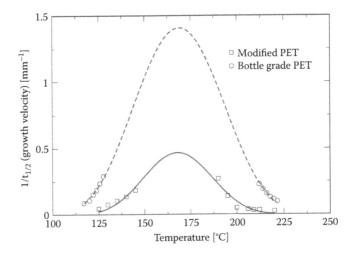

**FIGURE 9.3** Crystallization kinetics of unmodified and modified PET. (From Sorrentino L. et al., 2005, *Journal of Polymer Science Part B: Polymer Physics* 43, 15, August 1, 1966–1972, doi:10.1002/polb.20480. http://doi.wiley.com/10.1002/polb.20480.)

of polymer/platelet-like nanocomposites is one of the latest evolutionary steps of the polymer technology. Since the possibility of direct melt intercalation was first demonstrated, the melt intercalation method has become a main route for the preparation of the intercalated polymer nanocomposites without *in situ* intercalative polymerization. In recent years, intensive studies have been devoted to the complete exfoliation of the stacked silicate layers in the polymer matrix to develop materials with better properties (Figure 9.4). The use of nanoparticles can lead to new materials to be used for foaming since this will allow the modification of properties such as the gas absorption and diffusivity, local thermodynamic properties responsible for nucleation and growing phenomena, and rheological characteristics related to the final morphology of the foams.

Among the most important properties of a polymer-gas mixture in determining the foam structure are gas solubility and diffusivity in the molten polymer, mixture viscosity and viscoelastic behavior, and surface tension and transition temperatures (glass transition temperature and crystallization temperature). All these factors are significantly affected by the nature of components, by the concentration of the foaming agent and, obviously, by operating temperatures and pressures (see, for example, the works of Arefmanesh and Advani [1991] and Goel and Beckman [1995]). Investigation of solubility and diffusivity of polymer-gas mixtures is, hence, of primary importance. In particular, gas solubility determines the extent of the plasticization effect on the polymer and the final density of the foam and is one of the primary concerns when choosing physical blowing agents in a foam extrusion process (Pontiff, 2000).

Another key factor for foam processing is the mutual diffusivity of the blowing agent and the polymer. It affects the nucleation (nucleation has been described as a short range diffusion process) and the growth phenomena (gas has to diffuse out of the solution to inflate the bubbles), while from the processing point of view, diffusion

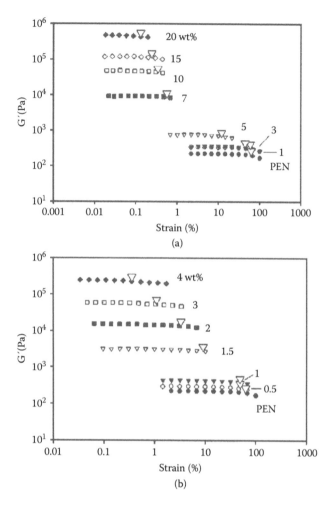

**FIGURE 9.4** Effect of nanoparticle dispersion on the viscosity of PEN nanocomposites based on (a) graphite oxide and (b) graphene. (Reported from Kim, H. and Macosko C. W., 2008, *Macromolecules* 41, 9, May, 3317–3327, doi:10.1021/ma702385h, http://pubs.acs.org/doi/abs/10.1021/ma702385h.)

fixes the residence time and the cooling rates on the extruded foam needed to avoid cell collapse. Mass transport properties of polymers can be described by using free volume theories (Vrentas and Duda, 1977), according to which mass transport is controlled by the availability of free volume within the system.

## 9.3  FOAMING TECHNOLOGIES FOR HPTP FOAMING

HPTP foams can be produced by using continuous (extrusion) or discontinuous technologies (injection molding and batch foaming). Each technique should be used according to the final article shapes and to the number of articles to be produced, but if a very fine morphology is required the batch foaming technique gives the best results.

### 9.3.1 Batch Foaming

The batch foaming process has, in the industrial practice, very little usage, since it is characterized by a very low productivity. High-added value products are manufactured using this technique (Harris, Kim, and Mooney, 1998; Krause, Mettinkhof et al., 2001; Mooney et al., 1996; Sorrentino, Aurilia, and Iannace, 2011; Sorrentino et al., 2011) such as the case of high-performance thermoplastics in biomedical or pharmaceutical fields. From a research point of view, instead, it has proved to be very useful, since it allows a fine control and a wide investigation range of the foaming variables and it is quite inexpensive.

The batch foaming technology is based on a thermo-regulated pressure vessel, whose temperature is controlled by means of an electric resistance or circulated oil bath. A schematic of the batch foaming apparatus is reported in Figure 9.5. The polymer is placed in the vessel and tightly enclosed. The blowing agent (usually a physical one) is then injected into the vessel at the desired pressure, and then the temperature is increased to a specified value (sorption temperature), to be defined according to the specific thermal and sorption properties of the selected HPTP. The physical blowing agent, PBA, diffuses into the polymer and after a defined time interval, depending on the transport properties of the gas/polymer mixture, saturates the sample. At the end of the saturation phase, a stable homogeneous solution is obtained. In order to foam the polymer, two procedures can be adopted: (a) temperature

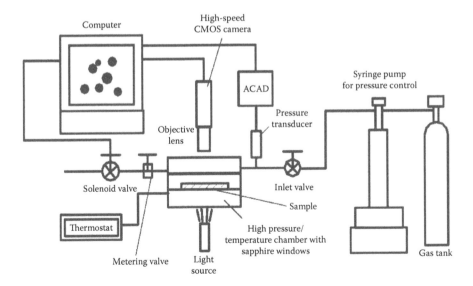

**FIGURE 9.5** Schematics of the batch foaming apparatus. (Reported from Guo Q. J. et al., 2006, *Industrial & Engineering Chemistry Research* 45, 18, August, 6153–6161, doi:10.1021/ ie060105w, http://pubs.acs.org/doi/abs/10.1021/ie060105w.)

increase and (b) pressure reduction. Typically, the former is used for glassy polymers (or quenchable, slow crystallizing semicrystalline polymers), while the second one for semicrystalline polymers.

### 9.3.1.1 Temperature Increase Foaming

Foaming by increasing the temperature is widely used to produce foams characterized by very fine structures (Krause, Mettinkhof et al., 2001; Krause et al., 2002) and they are mostly characterized by closed-cell structures. The polymer/PBA solution is to be cooled to a temperature well below the $T_g$ of the solution (well below the $T_g$ of the pure polymer) (Baldwin, Shimbo, and Suh, 1995; Goel and Beckman, 1993; Paterson et al., 1999; Paul et al., 1999). The pressure is then released from the pressure vessel and samples are immersed in a high temperature bath and maintained at that desired temperature (foaming temperature) during a predetermined foaming time. When the foaming temperature is higher than the glass transition temperature of the saturated polymer, the polymer expands and a porous structure is obtained. The porous samples are finally quenched to room temperature.

The influence of the processing parameters on the foam morphology and density was thoroughly studied by Krause, Mettinkhof et al. (2001), who showed that the most critical parameters are the saturation pressure and the foaming temperature. In the polysulfone (PSU)/carbon dioxide system, the cell density (number of cells per unit volume) increased exponentially with the increase of the saturation pressure, while the mean cell diameter decreased linearly. They also found that a characteristic foaming temperature range exists for each saturation pressure. Below the lower bound temperature (the glass transition temperature of the polymer/PBA mixture) the polymer/gas mixture is in a glassy state, which hinders the expansion of the polymer. Above the upper bound temperature (the glass transition temperature of the pure polymer), the nucleated cells tend to disappear because the polymer is not able to stabilize the cellular structure and the gas goes outside the cells. Kumar and Weller (1994) and Handa and Zhang (2000) performed similar investigations for polycarbonate (PC) and for polymethylmethacrylate (PMMA), respectively. They found comparable behaviors between the cell size and cell density patterns with respect to both dissolved carbon dioxide concentration and foaming temperature. These authors related the upper bound of the foaming temperature interval to cell coalescence: at higher temperatures the low elongational viscosity reduces the elongational stresses bearable by cell walls, in turn causing their rupture and the coalescence of the cellular structure. The exact value of the upper bound temperature depends on the polymer properties, namely its melt viscosity, melt strength, surface properties, and transport properties of the polymer/PBA mixture.

### 9.3.1.2 Foaming by Pressure Quenching

This technology is used when the polymer is capable of fast forming a crystalline phase that can hinder the cell collapsing after expansion at foaming temperature. The saturation temperature has to be higher than the melting temperature of the pure polymer, in order to melt the crystalline phase and to allow the blowing agent (usually of physical type) solubilization in the whole mass of polymer. After solubilization, the mixture is cooled down to the foaming temperature and then pressure quenched to atmospheric pressure for foaming. Finally, the temperature is reduced to room temperature. The pressure reduction is needed to induce the thermodynamic instability responsible for the supersaturation of the blowing agent in the mixture and for its subsequent migration into the nucleated bubbles. Also, in this case the blowing agent diffusion from the polymer reduces the plasticization of the polymer matrix and increases the transition temperatures ($T_g$ and $T_m$) of the system, contributing to the stabilization of the developed cellular structure.

The density of the foam decreases with the increase of the saturation concentration, as more blowing agent is available for cell growth. Furthermore, the increase of blowing agent amount in the polymer determines an increase of the number of cells (cell density) because of a more intense nucleation. The foaming temperature has to be selected in order to avoid the premature crystallization of the foaming solution but has to be low enough to allow cellular structure consolidation after foaming, either for polymer crystallization or viscosity increase. The useful temperature range is quite narrow with this technology because low temperatures lead to the formation of foams with good morphology but high density, while high temperatures lead to low density foams but large diameters. If the foaming temperature is too high, cell collapse occurs.

Pressure drop rate is also very important because nucleation and growth compete in subtracting the expanding gas from the molten polymer/gas solution and if growth is allowed no further nucleation can take place around the growing bubble.

## 9.3.2 EXTRUSION FOAMING

Extrusion is the leading technology in thermoplastic processing and extrusion foaming is the most important foaming process. In fact, it is continuous (a must have for high throughput production), involves a small amount of material (very important for effective temperature control, because the foaming processing window for thermoplastics is typically narrow), and is able to produce both finished or semifinished, for further processing, products (Klempner and Frish, 1991; Lee, 2000). Extrusion foaming process is more critical for HPTP polymers with respect to commodity polymers because all the thermal phenomena involved in the foaming process are radicalized as a result of the typical abrupt temperature drop at the die exit. It is thus fundamental to properly account for or to increase the crystallization kinetics in semicrystalline polymers to stabilize the cellular structure, or to adequately manage the glass transition temperature reduction for the plasticizing effect of the blowing agent in amorphous polymers.

Furthermore, extrusion allows the use of both chemical and physical blowing agents, which need specific treatments to be correctly handled. The most critical operations for producing thermoplastic foams by extrusion are summarized in the following.

### 9.3.2.1 Feeding and Melting

A proper feeding of all of the components is crucial to obtain stable foam. In fact, whether chemical (CBA), added through the feeder, or physical (PBA), to be injected along the extruder, blowing agents are used, the initial composition fixes the final density of the foam. Additives such as nucleating agents have a strong effect on cell nucleation and, therefore, on the final foam morphology. Care must be taken in setting the barrel temperature to avoid the premature decomposition of the blowing agent, if a CBA is used.

In the first path of the extruder, the polymeric material is heated above the melting temperature and compounded to the additives. This second task is not trivial as a fine dispersion of the additives, in particular the nucleating agents, is necessary for obtaining a stable and a high-quality foam. The nucleating agents tend to agglomerate in clusters, the fracture of which often justifies the use of twin-screw extruders. This stage immediately precedes the gas injection or the CBA decomposition zone and melting has to be completed in this section of the extruder because if solid polymer is still present, the blowing agent (from the injection port or the CBA decomposition zone) can escape through the hopper. In order to avoid this issue, a dynamic sealing can be used.

### 9.3.2.2   PBA Injection/CBA Decomposition

The gas or liquid blowing agent can be injected at a pressure higher than that in the barrel at injection point. The mass flow rate has to be constant, as it determines the final density of the foam but also the plasticizing effect on the molten polymer, which has a dramatic effect on the processability of the expanding mixture. If a CBA is used it must not prematurely decompose, and temperature should be adequately increased to promote the CBA decomposition. The failure to reach the decomposition temperature, which could result in inadequate density reduction, or the premature CBA decomposition is faced by using activating agents that decrease the decomposition temperature or retard the decomposition.

### 9.3.2.3   Mixing and Solubilizing

Dissolution of the blowing agent to form a single phase with the molten polymer is of primary importance. The solubilization is limited by the equilibrium gas concentration at the processing pressures and temperatures, and it limits the lowest foam density attainable. It is very important to reduce the solubilization time, usually by means of a temperature increase or an intense mixing, to reduce the residence time. The temperature can be set to a value below the melting point of the polymer, as the plasticizing effect of the blowing agent lowers the polymer/gas solution melting temperature.

### 9.3.2.4   Cooling and Die Forming

The most peculiar phase of the extrusion foaming is the cooling of the polymer/gas solution prior to die exit. It is also very critical, since a bad temperature control determines unsuccessful foaming or polymer premature crystallization. Temperature gradients should be avoided, to prevent underblown or overblown portion of the extrudate, by means of optimal screw design, a proper temperature control, and the use of static mixers.

In the foaming process, the exit die has the very important role of controlling the pressure drop responsible for bubble formation and the pressure drop distribution along the die to prevent premature bubble formation. The die temperature, which corresponds to the foaming temperature of the polymer/gas solution, must be properly defined to reach the desired foam density and to avoid cell collapsing. Thanks to the plasticizing effect of the gas, the die temperature can be set to a value below the crystallization temperature of the neat polymer, in turn also exploiting the viscosity increase during gas escaping from the matrix to further consolidate the cellular morphology developed.

### 9.3.3 INJECTION MOLDING

Another important processing technique for foaming is the injection molding, which is a modification of the conventional injection-molding process. As with the extrusion process, the selection of the blowing agent type (chemical or physical) influences the process, in terms of the way of performing the injection of the polymeric solution into the mold.

The injection-molding process combines the extrusion process with the injection one through the use of alternating movement of the screws. This alternative movement causes pressure oscillations that have to be controlled in order to avoid premature nucleation of the bubbles in the barrel, in particular when the polymeric solution approaches the last zone of the barrel. The pressure must be kept above the saturation pressure by using the backpressure, which is the pressure applied during the backward movement of the screws.

After the die exit, the solution is injected into the mold and is expanded by the blowing agent. The injection rate and the mold design determine the nucleation kinetic and growth. The faster the injection, the higher the pressure drop and, as a consequence, the smaller the bubbles. Such as in the extrusion foaming process, the amount of foaming agent influences both the mean cell size and the density of the molded object.

The very high temperatures to be used with HPTPs also in this case radicalize the thermal issues, in particular the interaction of the cell growth process and the consolidation (during cooling in the mold) of the cellular structure. In fact, as the foaming melt solution is injected into the mold cavity, a portion of the polymer touches the cold mold walls, thus quenching and hindering the expansion. A layer of solid polymer develops along the mold wall, whose thickness should be minimized through properly selecting the injection conditions, imparting a skin/core structure to the article. Before opening the mold, the molded part has to be completely solidified, otherwise warping or volume change will occur. This could be a serious issue due to the very low thermal conductivity of the foam and also leads to a low productivity or polymer degradation in the extruder barrel. The plasticizing effect of the blowing agent on the polymer can help because it allows the lowering of the injection temperature with respect to the bulk polymer.

## 9.4 HIGH-PERFORMANCE THERMOPLASTICS AND THEIR FOAMS

### 9.4.1 AROMATIC POLYKETONES

High-performance semicrystalline thermoplastic polymers such as aromatic polyetherketones are ideal candidates as matrices in nanocomposite materials due to their unique combination of high glass transition temperature ($T_g$), stiffness, strength, toughness, chemical and solvent resistance, thermo-oxidative stability, flame retardancy, low dielectric constant, retained properties at very high temperatures, long storage periods, and recyclability (Díez-Pascual, Naffakh, Marco, Ellis et al., 2012). Aromatic polyketones (PEEK, PEK, PEKK, PAEK) are widely employed in aeronautics and high-end applications due to their high stiffness and strength, high resistance to a huge number of solvents, for example, Jetfuel, Skydrol, and methylethylketone (MEK), and have a good flame retardancy. They are crystalline thermoplastic polymers and have a glass transition temperature ranging between 140 and 165°C, and a melting temperature ranging between 340 and 375°C. They exhibit very good mechanical properties at room temperature, such as an elastic modulus up to 4.5 GPa (PEKK) and a tensile strength of 110 MPa.

Polyetherketone-based nanocomposites can be produced by conventional polymer processing techniques such as extrusion and compression molding, frequently combined with preprocessing stages such as mechanochemical treatments in organic solvents. In order to further increase their properties, nanofiller addition has been considered but dispersion and improved load transfer ability from the matrix to the nanoreinforcement must be achieved and different routes have been developed to efficiently incorporate nanofillers into these matrices.

A wide variety of nanofillers have been investigated. In the case of carbon nanofillers polymer functionalization (Blundell and Osborn, 1983; Díez-Pascual, Martínez, and Gómez, 2009), covalent grafting (Oh et al., 2006), and nanofiller wrapping in compatibilizing systems (Díez-Pascual et al., 2009; Díez-Pascual et al., 2010) have been tried to improve the performance of polyketone-based nanocomposites (Díez-Pascual, Naffakh, Marco, Ellis et al., 2012).

Mishra et al. (2012) prepared a PEEK nanocomposite by using zirconia nanoparticles observing that their presence enhanced various basic and functional properties,

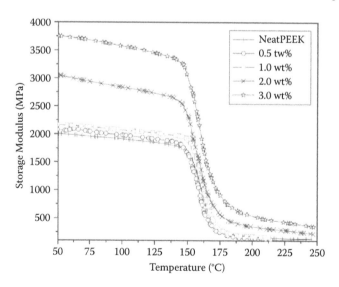

**FIGURE 9.6** Storage moduli of zirconia/PEEK nanocomposites. (Reported from Mishra T. K. et al., 2012, *Composites Science and Technology* 72, 13, August, 1627–1631, doi:10.1016/j. compscitech.2012.06.019, http://linkinghub.elsevier.com/retrieve/pii/S0266353812002436.)

in particular the dynamic mechanical ones (Figure 9.6). A uniform dispersion of nanoparticles was achieved by using melt blending in a twin-screw extruder for proper dispersion of the silane-modified zirconia powder.

Nanoreinforced fibers based on PEEK containing up to 10 wt% vapor-grown carbon nanofibers were produced by using conventional polymer processing equipment such as a twin-screw extruder (Sandler et al., 2003) or injection molding (Verdejo et al., 2009). Mechanical tensile testing revealed increases in nanocomposite stiffness, yield stress, and fracture strength and also the alignment of nanofibers along the direction of flow during processing. Furthermore, the degree of crystallinity of the poly(ether ether ketone) matrix increased with the addition of nanofibers. MWNTs/PEEK nanocomposites (Deng, Ogasawara, and Takeda, 2007), nano-zinc sulfide (ZnS)- and nano-titanium dioxide ($TiO_2$)-reinforced PEEK composites (Knör et al., 2009), nano-$SiO_2$/PEEK (Kuo et al., 2010), nano-$SiO_2$/short carbon fiber PEEK ternary system (Zhang, Chang, and Schlarb, 2009) PEEK (Balaji, Tiwari, and Goyal, 2011), PEEK/(aluminum nitride) AlN; (Goyal, 2009), PEEK/inorganic fullerene-like $WS_2$ (Naffakh et al., 2010), Nano-$TiO_2$/PEEK (Wu et al., 2012), and MMT (Cloisite 15A)/PEEK (Jaafar, Ismail, and Matsuura, 2012) were successfully produced by conventional plastics techniques (mainly twin-screw extrusion) and evidenced good dispersions and improvements in mechanical as well as functional (thermal, dynamic mechanical) properties. Also, the use of a compatibilizer was explored to enhance the dynamic mechanical and thermal properties of PEEK/carbon nanotube systems (Díez-Pascual et al., 2009).

Polyketones have also been investigated with respect to their foamability, but only limited investigations were performed on nanocomposite systems. Some analysis of the sorption properties and crystallization behavior of carbon dioxide in PEEK have been reported by Wang et al. (2006, 2007b).

Microcellular PEEK foams have been prepared by using a two-stage batch foaming process by using carbon dioxide as blowing agent. Two cell sizes were evident in samples (Figure 9.7), exhibiting a sandwich-like structure with nucleated cells and cell size depending on the blowing agent uptake and on viscoelastic properties, respectively (Wang et al., 2007a).

Microcellular foams were also produced by Behrendt et al. (2006), who expanded thin PEEK films by using the batch foaming technique with the aim to prepare electrets. They obtained a low porosity structure, as a consequence of the crystallization during the foaming process, but the polymer was crystallized after the foaming process. Blends based on PEEK have been used to exploit the interfaces as nucleating sites in Tan et al. (2003) and Nemoto, Takagi, and Ohshima (2010). In particular, Nemoto et al. were able to obtain nanoscale cell structure from poly(ether ether ketone) (PEEK)/para-diamine poly(ether imide) (p-PEI) as well as PEEK/meta-diamine poly(ether imide) (m-PEI) blends by a temperature quench foaming process with $CO_2$ as blowing agent. They stated that the difference in chemical configuration between m- and p-PEI gave rise to a prominent change in the higher-order blend morphology and cell structure of the respective foams. The bubble nucleation site and bubble size were controlled by templating the morphology of the PEEK/p-PEI blend, which showed an immiscible and unique strip-patterned crystalline morphology (Figure 9.8). The properties influenced by the immiscibility of the PEEK/p-PEI blend, investigated using SEM, thermal analysis, and rheology, and compared with

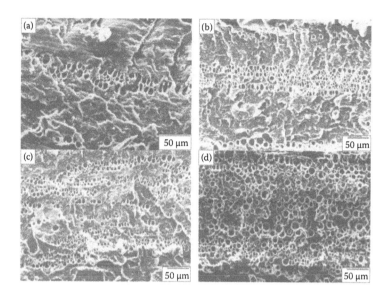

**FIGURE 9.7**   SEM images of foaming films with different sorption times: (a): 40 s; (b): 5 min; (c): 15 min; (d): 30 min (foaming temperature: 140°C; foaming time: 10 min). (Reported from Wang D. et al., 2007a, *Journal of Polymer Science Part B: Polymer Physics* 45, 20, October 15, 2890–2898, doi:10.1002/polb.21266, http://doi.wiley.com/10.1002/polb.21266.)

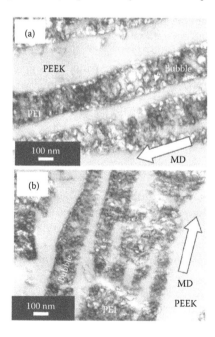

**FIGURE 9.8**   TEM micrographs of a cross-sectional area of (a) p-PEI-40 and (b) p-PEI-60-cast film foamed at 200°C. (MD: machine direction, to which the film was stretched.) (Reported from Nemoto T. et al., 2010, *Polymer Engineering & Science* 50, 12, December 30, 2408–2416, doi:10.1002/pen.21766, http://onlinelibrary.wiley.com/doi/10.1002/pen.21766/full.)

the properties of the miscible PEEK/m-PEI blend, allowed understanding that the bubble size and location were highly controlled in the PEI disperse domain that was aligned between the PEEK crystalline layers in the PEEK/p-PEI blend (Nemoto, Takagi, and Ohshima, 2010).

Attempts at producing foams on PEEK-based nanocomposites have been performed by Verdejo et al. who used vapor-grown carbon nanofibers (CNFs) to stabilize the foaming process through an injection-molding process, in which they used commercial chemical blowing agents. They produced integral foams with both unfilled and nanocomposite systems and verified the positive effect of the nanofiller on the cellular structure (Figure 9.9). A homogeneous dispersion of the nanofibers in the cellular PEEK cores was achieved, which was responsible for the mechanical properties increase, in bending, of the foam injection-molded samples with nanofiber loading fractions up to 15 wt%, demonstrating a strong interaction between the matrix and the nanoscale filler (Verdejo et al., 2009).

Macroporous foams based on PEEK were also produced (Figure 9.10). Particle-stabilized liquid foams provide a general route for producing low-density macroporous materials from either melt-processable or intractable thermoplastics. By varying the size, concentration, and wettability of the particles in the colloidal suspensions

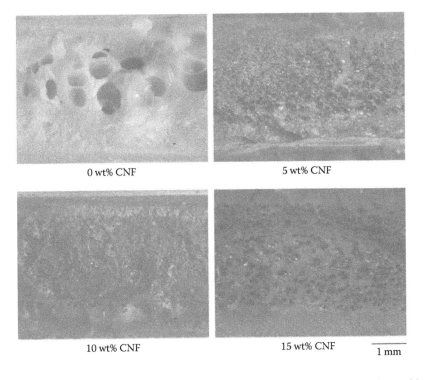

0 wt% CNF                5 wt% CNF

10 wt% CNF               15 wt% CNF              1 mm

**FIGURE 9.9** Optical light microscopy images of the fracture surfaces of injection-molded foam samples produced using a commercial chemical blowing agent. (Reported from Verdejo et al., 2009, *Journal of Materials Science* 44, 6, January 15, 1427–1434, doi:10.1007/s10853-008-3168-y, http://www.springerlink.com/index/10.1007/s10853-008-3168-y.)

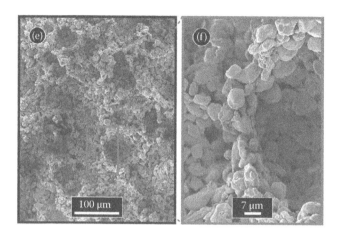

**FIGURE 9.10**   Particle stabilized foams of PEEK from liquid solution. (Reported from Akartuna et al., 2009, *Polymer* 50, 15, July, 3645–3651, doi:10.1016/j.polymer.2009.05.023, http://linkinghub.elsevier.com/retrieve/pii/S0032386109004297.)

and controlling the processing conditions, macroporous materials with porosity of 74% and a median size of particles between 80 and 180 μm were obtained (Akartuna et al., 2009; Wong et al., 2010).

### 9.4.2   Polyesters (PET, PEN)

Among the most diffuse and commercially available polyesters, PEN is the only one that has adequate performances for the inclusion in the HPTPs class, because it has a glass transition temperature of 125°C and a melting temperature higher than 265°C. PET, although commonly included within engineering polymers because of its glass transition temperature ($T_g = 75$°C), could be included within HPTPs if conveniently reinforced with nanoparticles, which are able to enhance its mechanical as well as functional characteristics.

Several nanoparticles of different shape factors have been considered for PET reinforcement. Platelet-like nanoparticles, such as MMT or graphene, have demonstrated allowance of a strong enhancement of heat deflection and glass transition temperatures, of elastic modulus and dynamic mechanical performances. Clay nanoparticles improve barrier properties to gas sorption (Hayrapetyan et al., 2012) and are also capable of enhancing crystallization kinetics of the polymer (Durmus et al., 2009). This is a very important issue for cellular morphology stabilization since crystallization kinetics of foaming grade PET is slowed by the molecular weight increase after the chain extension process (Sorrentino et al., 2005). PET nanocomposites have been produced through different routes, but melt blending techniques are preferred due to the ease of the process and its productivity (Ghasemi et al., 2011; Özen et al., 2012; Tsai et al., 2005). Since the organic modifier in conventional clay nanoparticles is not able to bear high processing temperature, more thermally stable modifiers have been investigated to allow a good intercalation and to limit polymer degradation during extrusion (Scamardella et al., 2012).

Carbon-based nanofillers do not show thermal degradation issues, and thermal and thermo-oxidative degradation temperatures of PET/EG nanocomposites are improved substantially with the increment of their content. Graphene (Zhang et al., 2010) as well as expanded graphite (Li and Jeong, 2011), have been successfully melt blended into PET to increase the mechanical (both static and dynamic mechanical) and functional (thermal stability, electrical conductivity) properties. Carbon nanotubes have been also used as reinforcing nanophase into PET matrix and have shown similar performance improvement of graphite nanoparticles (Aurilia, Sorrentino, and Iannace, 2012; Kim, Choi et al., 2009; May-Pat, 2011; Yesil and Bayram, 2011). Nanopowder-based nanocomposites did not show significant mechanical performance improvement, but were able to increase the crystallization properties of PET nanocomposite, such as in the case of silica nanoparticles (Bikiaris, Karavelidis, and Karayannidis, 2006; Liu et al., 2004).

PET nanocomposites have been widely investigated from the sorption (Baldwin, Shimbo, and Suh, 1995; Pavel and Shanks, 2003) and crystallization kinetics (Ghasemi, Carreau, and Kamal, 2012; Ke, Long, and Qi, 1999; Sorrentino et al., 2005) points of view and different routes have been explored to increase their molecular weight in order to get foamability. To increase rheological properties, reactive extrusion is the main technique used to bond polymer end groups through the use of multifunctional epoxy-based additives or pyromellitic dianhydride (Japon et al., 2000; Nguyen et al., 2001; Xanthos et al., 2001, 2004), even if postcondensation has been considered because of its lower processing temperatures and its higher control on the chain extension reactions with respect to the reactive extrusion (Al Ghatta and Cobror, 1998).

Studies on mechanisms and processing for the production of closed-cell microcellular foams were performed in the 1990s by Baldwin et al., who used solid-state foaming technique to induce foaming in a carbon dioxide solubilized quenched PET film (Baldwin, Park, and Suh, 1996). Despite the large number of works on the foaming of neat PET, only limited investigations are available on PET-based nanocomposite matrices, due to the difficulties to stabilize the cellular structure after the melt blending process. In 2012, Scamardella et al. used a new approach to compensate the thermal degradation of both polymer macromolecules and inorganic modifier at the high processing temperatures with the use of pyromellitic dianhydride directly during the nanofiller melt blending (Scamardella et al., 2012). The nanofiller was used to increase the crystallization kinetics, to improve the cell's nucleation and cell's stabilization during the batch foaming process, and to extend the maximum working temperature of the foams as a result of the gain in polymer viscosity related to the good platelet dispersion.

Regarding PEN nanocomposites, $SiO_2$-based systems have been developed since 2004 by Ahn, Kim, and Lee (2004), who surface-modified silica nanoparticles to improve their dispersion in the polymer during melt blending in an internal mixer. They reported on the increase of the elastic modulus and elongation at break at low nanoparticle content (0.4%wt). On the same nanocomposite system, Kim et al. showed an increase of several chemical-physical properties (thermal stability, dynamic mechanical behavior, and crystallization kinetics) after nanoparticle addition, and also a fourfold gain in the elastic modulus above the glass transition temperature with 2.0% by weight of silica nanoparticles (Kim, Kim, and Kim, 2009).

Carbon nanotubes have been widely investigated as a PEN nanofiller. Kim et al. obtained a huge improvement of the viscosity by adding 2.0%wt of MWCNT, but they obtained a plateau of both tensile modulus and tensile strength after the addition of just 0.5%wt of nanofiller. This behavior could be attributed to the difficulties of dispersing a high amount of carbon nanotubes and to the presence of ineffective particles aggregates (Kim and Kim, 2006). Further developments also showed increases in the crystallization kinetics (Kim, Han, and Kim, 2007; Kim, Han et al., 2009), thermal stability, and elastic modulus above the glass transition temperature with respect to the neat polymer (Kim, Kim, and Kim, 2009).

Layered silicate nanoparticles have also been used to prepare PEN-based nanocomposites through the direct intercalation of PEN polymer chains from the melt into the surface-treated clay. An internal mixer was used and exfoliated silicate layers within a PEN matrix were obtained. Mechanical and barrier properties measured by dynamic mechanical and permeability analysis showed significant improvements in the storage modulus and water permeability when compared to neat PEN (Wu and Liu, 2005).

In order to limit degradation phenomena related to the organic modifier present as intercalant of platelets, exfoliated graphite has been considered by Kim and Macosko (2008). They obtained a very good dispersion of graphene sheets, by means of a twin-screw extruder, which was responsible for the significant improvements in viscosity, electrical conductivity, gas barrier, and mechanical properties. The use of exfoliated graphite in nanocomposites exhibited strong improvements with respect to microsized graphite composites. Expanded graphite was used by Sorrentino et al. (2010) to prepare PEN nanocomposites as matrix for foamed materials. They reported on the improved crystallization kinetics of nanoparticles, coupled with the nucleating activity for both crystals and cells.

Closed-cell PEN foams were for the first time produced by Sorrentino, Aurilia, and Iannace in 2011 (2011). They used a batch foaming apparatus to solubilize carbon dioxide into amorphous PEN. Foams were produced within an extended temperature range (from 100 to 240°C) and were characterized by relative densities ranging from 0.13 to 0.44 and average cell size between 5 and 15 μm. Cell nucleation strongly increased above PEN glass transition temperature, as a consequence of the higher molecular mobility.

Expanded graphite (EG) and silica nanocomposites based on PEN were prepared by means of a twin-screw extruder by Sorrentino et al. (2011a). The presence of platelet-like reinforcement (EG) reduced the carbon dioxide uptake, while silica nanoparticles only slightly affected the sorption properties. EG nanoparticles were also more effective as a nucleating agent for both crystallization and foaming with respect to silica nanoparticles (Sorrentino et al., 2011b). In fact, $10^{10}$ cells per $cm^3$ were obtained with EG, while neat PEN and $SiO_2$ nanocomposite only showed $10^9$ cells per $cm^3$. Foam density resulted as low as 0.15 and 0.11 $g/cm^3$, for EG and $SiO_2$ nanoparticles, respectively, but a microcellular morphology was only obtained after EG addition (Figure 9.11). Planar nanoparticles also resulted in the increase of the compressive modulus in foams, thanks to the reinforcing effect of the stiffer EG platelets, which aligned along cell walls during cell growth (Sorrentino et al., 2012).

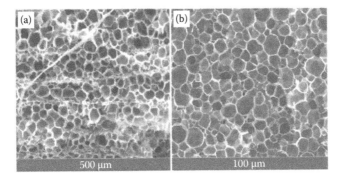

**FIGURE 9.11** Comparison of the cellular morphology of foams from (a) neat PEN and (b) 1.0%wt EG/PEN nanocomposite. (Reported from Sorrentino L. et al., 2012, *Journal of Cellular Plastics* 48, 4, July 12, 355–368, doi:10.1177/0021955X12449641, http://cel.sagepub.com/cgi/content/abstract/48/4/355.)

### 9.4.3 SULFONE POLYMERS

Sulfone polymers are amorphous high $T_g$ polymers, which are characterized by resistance up to 220°C, good mechanical properties and high fracture resistance, resistance to superheated steam, and exceptional resistance to chemicals. These polymers are employed in several applications in construction of equipment, in the electrical and electronics fields, and in automotive engineering. Recently, food industry has discovered these polymers because they can be available in FDA (United States) compliant grades. They have long-term service temperatures of up to 190°C, good dimensional stability, creep strength at high temperatures, and are characterized by good resistance to oil, fuels, and fluorine even if they are in the amorphous state.

Sulfone polymers have been widely investigated since 2009 as a matrix for nanocomposites. Li et al. used $TiO_2$ nanoparticles to improve the performance of microporous polyethersulfone (PES) membranes (Li et al., 2009). Aurilia et al. investigated the effects of $SiO_2$ and expanded graphite (EG) nanoparticles on the solvent sorption and mechanical properties of PES nanocomposites. Both nanoparticles improved the solvent resistance of the amorphous matrix acting as a barrier to the solvent diffusion in the polymer (EG to a higher extent), but only EG induced a strong improvement on the structural response (Aurilia et al., 2010).

PES nanocomposites have been investigated to increase the filtration performance of membranes, by incorporation of carbon nanotubes (Park et al., 2010), organically modified clay (Ghaemi et al., 2011; Liang et al., 2012), $TiO_2$ (Liang et al., 2012; Madaeni et al., 2012), modified halloysite nanotubes (Zhang et al., 2012), $SiO_2$ nanopowders (Huang et al., 2012), and multi-wall carbon nanotubes (Yun et al., 2012).

Foams from sulfone polymers have been prepared since 2001. Krause et al. investigated the foaming process of poly(ether sulfone) (PES) and polysulfone (PSU), by using carbon dioxide as a physical blowing agent and the temperature rise foaming technique after gas solubilization. They measured a strong glass transition temperature depression after gas sorption (Figure 9.12), and a wider temperature range for foaming was detected at high-solubilization pressures (Figure 9.13). Microcellular

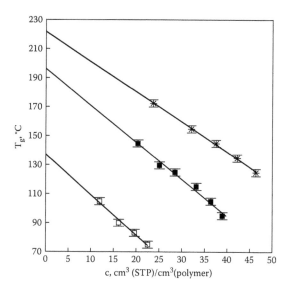

**FIGURE 9.12**  Glass transition temperature of PSU (■) and PES (*) dependent on the dissolved amount of carbon dioxide. (Reported from Krause B. et al., 2001, *Macromolecules* 34, 4, February, 874–884, doi:10.1021/ma001291z, http://pubs.acs.org/doi/abs/10.1021/ma00 1291z.)

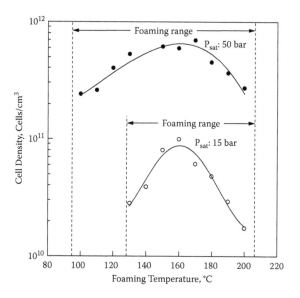

**FIGURE 9.13**  Cell density of PSU for different foaming temperatures saturated with 15 bar and 50 bar carbon dioxide for 2 h at room temperature. Foaming times of 30 s were used. The dotted lines confine the foaming temperature range in which foam formation is observed. (Reported from Krause B. et al., 2001, *Macromolecules* 34, 4, February, 874–884, doi:10.1021/ma001291z, http://pubs.acs.org/doi/abs/10.1021/ma001291z.)

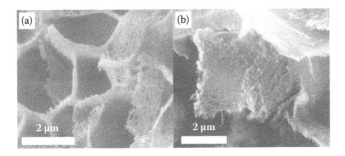

**FIGURE 9.14**    SEM micrographs of (a) PES and (b) PEI foams showing the sub-microcellular structures with nanoporous walls produced at 240°C. (Reported from Sorrentino, L., M. Aurilia, and S. Iannace, 2011, *Advances in Polymer Technology* 30, 3, September 8, 234–243, doi:10.1002/adv.20219, http://onlinelibrary.wiley.com/doi/10.1002/adv.20219/full.)

closed-cell morphologies were obtained, but foam density was higher than $0.4 \text{g/cm}^3$ (Krause, Mettinkhof et al., 2001).

Blends of PES and PEI were also investigated to exploit the combined action of glass transition temperature depression after blowing agent sorption and to the interfaces between the different polymer matrices. The result was the production of foams with nanocellular cells, but very high foam density (Krause, Sijbesma et al., 2001).

Investigations on the mechanical response of PES and PSU foams was performed by Sun et al. who used a two stage batch process to induce a microcellular morphology (Sun and Mark, 2002; Sun, Sur, and Mark, 2002).

Poly(phenyl sulfone) (PPSU) have also been investigated for foaming by Sorrentino, Aurilia, and Iannace (2011). They detected a reduced foamability with respect to PES, showing lower nucleated cells, higher relative foam density (0.4 and 0.2 for PPSU and PES, respectively), and average cell size (9 µm and 1 µm for PPSU and PES, respectively). The coupled effect of $CO_2$ plasticization and fast cell formation lead to interesting cell wall morphologies, exhibiting nanometrix voids as shown in Figure 9.14.

Van Houten and Baird produced PAES foams with a pressure quench technique, using both carbon dioxide ($CO_2$) and water as well as nitrogen and water as physical blowing agents. Foams produced with a water/$CO_2$ system yielded foams with better properties than nitrogen and water because both water and $CO_2$ are plasticizers for the PAES resin. A cell size of ~50 µm, while maintaining a primarily closed-cell structure, was obtained, thus enhancing the mechanical properties of the foam (Van Houten and Baird, 2009). The further addition of the CNFs greatly reduced the mean cell size, increased the nucleation rate and the compressive properties while slightly decreasing the impact strength of the foams. By adding in only 1% of the CNFs, both the specific compressive modulus and specific compressive strength were increased by over 1.5 times the modulus and strength of the unreinforced PAES foam.

PES nanocomposites prepared with $SiO_2$ and EG nanoparticles were studied by Sorrentino et al. They have found that the use of silica nanoparticles did not affect the solubility of carbon dioxide, while EG reduced it at high filler loadings. Foams were obtained with the temperature rise technique after solubilization in a high-pressure vessel. EG filled PES foams exhibited lower foam density, with respect to $SiO_2$ filled ones, and allowed a two order of magnitude increase of the nucleated cells up to $10^{11}$ cells per $cm^3$ ($SiO_2$ gave a one order of magnitude increase) with respect to unfilled matrix (Sorrentino et al., 2011).

### 9.4.4 Thermoplastic Polyimides (PI, PAI, and PEI)

Polymers with imide groups in their backbone possess very special properties that make them prime candidates when high temperature material performance is required. They are characterized by some of the highest glass transition temperatures available among thermoplastics, up to 285°C for poly(amide imide), depending on the aromatic groups in the repeating units. This valuable property is also the main weakness, from the processing point of view, since thermoplastic polyaryimide (PAI) and polyimide (PI) cannot be processed by melt compounding techniques, and methods such as solution casting, powder sintering, sol gel, or precursors curing have to be used. Only poly(ether imide) polymers, among PI class, have been processed by melt compounding at temperatures usually higher than 350°C because its glass transition temperature is around 220°C and viscosity should be lowered to acceptable values during extrusion or injection molding.

The preparation of polyimide-based nanocomposites is even more difficult because viscosity during nanoparticle dispersion strongly rises. For this reason, methods such as precursor curing, for the preparation of layered silicate nanocomposites (LeBaron, 1999), or sol gel (Chen and Iroh, 1999), for the dispersion of silica nanoparticles, have been used in the past. Other nanoparticles have been investigated to develop PI nanocomposites by using thermal imidization processes, such as $Al_2O_3$ (Cai et al., 2003), thermally stable aromatic amines-modified montmorillonites (Liang, 2003), dual intercalating agent system montmorillonites (Huang et al., 2008), and carbon nanotubes (Cai, Yan, and Xue, 2004).

Some investigations have been made on thermoplastic polyimide nanocomposites, but they were specifically developed in the laboratory. Golubeva et al. developed nanocomposites based on a thermoplastic polyimide, poly{1,3-bis(3',4-dicarboxyphenoxy) benzene[4,4'-bis(4''-N-phenoxy)diphenyl sulfone]imide}, and synthetic magnesium silicate nanoparticles with montmorillonite structure, prepared from melts after a modification of the nanoparticle surface with (aminoethylaminomethyl)phenethyltrimethoxysilane (Golubeva et al., 2007). Yudin and Svetlichnyi prepared graphene sheet-based nanocomposites using specially synthesized amorphous and semicrystalline polyimide matrices, and employed silicate (natural and synthetic) and carbon nanoparticles with different morphology (tubes, platelets, discs, and spheres), studying the influence of nanoparticle morphology on the rheological behavior of nanocomposite melts during processing (Yudin and Svetlichnyi, 2010). They were able to work with melt blending technologies, such as melt mixing and injection molding, to prepare blends and samples at temperatures between 350 and 360°C. PAI/MMT nanocomposites have been prepared since 2002 by means of solvent suspension technique to intercalate or exfoliate nanoparticles followed by thermal imidization to consolidate the polymer (Faghihi, Shabanian, and Bolhasani, 2012; Ranade, D'Souza, and Gnade, 2002). Sol gel technique, also followed by thermal imidization, has been used to disperse MMT (Shantalii et al., 2005), silica nanoparticles (Son et al., 2008), and multi-walled carbon nanotubes (Lee et al., 2010). In all cases, absorption properties of PAI/nanocomposites decreased significantly, while the thermal stability increased significantly with increasing the nanofiller content in PAI matrix.

Poly(etherimide) nanocomposites were developed in 2001, when Huang et al. prepared PEI/montmorillonite nanocomposites by melt intercalation of nanoparticles specifically modified with thermally stable aromatic amines (Huang et al., 2001). They used an internal mixer to disperse MMT nanoparticles, obtaining a very good particle dispersion (exfoliated structure), and also verified that nanocomposites exhibited a substantial increase of glass transition temperature and thermal decomposition temperature, and a dramatic decrease in solvent uptake compared to the virgin PEI due to the molecular dispersion of the MMT layers and the strong interaction between PEI and MMT (Liang and Yin, 2003).

Other nanofillers used in PEI nanocomposites are single-wall carbon nanotubes (Siochi et al., 2004), multi-walled carbon nanotubes (Liu, Tong, and Zhang, 2007), also prepared with ultrasound assisted twin-screw extrusion (Isayev, Kumar, and Lewis, 2009; Shepard et al., 2012), nano-TiO2, and short carbon fiber (through twin-screw extrusion) (Chang et al., 2005; Xian, Zhang, and Friedrich, 2006). Solution casting has also been used to disperse MWCNT (Shao et al., 2009), BaTiO$_3$ (Choudhury, 2010) or aluminum nitride nanoparticles (Wu et al., 2011).

Polyimide foams were prepared in 1995 by Hedrick et al. by using a blowing agent developed during thermal treatment of the polymer (Charlier et al., 1995; Hedrick, 1995; Hedrick et al., 1996, 1998). Foams with pore sizes in the nanometer regime were prepared by casting block copolymers comprising a thermally stable block and a thermally labile material, such that the obtained morphology was characterized by the matrix made of the thermally stable material and a dispersed phase made of the thermally labile material. Thanks to temperature rise, the thermally unstable block degrades by thermolysis, and leaves pores with size and shape defined by the initial copolymer morphology.

A different approach was used by Weiser et al. (2000), who developed a method where polyimide foams were made from powders of solid-state polyimide precursors (i.e., poly[amic acid], PAA) by using tetrahydrofuran (THF) as a blowing agent that complexes through hydrogen bonding to the poly(amic acid) structure. During the foaming process, solid powder particles of poly(amic acid) precursors are heated from room temperature to produce microspheres and ultimately foams. A deep understanding of this method was performed by Cano et al. (2005) who investigated the competitive diffusion phenomena, which is verified during the foaming process.

Only recently, microcellular structure was directly induced in large-sized polyimide foams. In fact, the preparation of lightweight microcellular polyimide foams with a large size is challenging and inefficient, because of low gas solubility, high stiffness, and an extremely long saturation time. A viable solution to these problems was the use of solid-state microcellular foaming technology with the compressed CO$_2$ as a physical blowing agent and tetrahydrofuran as a coblowing agent (Zhai et al., 2012). The presence of a coblowing agent increased the gas sorption of PI, causing a dramatic increase in the expansion ratio of microcellular PI. The authors also verified the possibility of preparing PI foamed beads to employ compression molding processes to 3D shape products. They coated beads with a poly(ether imide) (PEI)/chloroform solution, in order to well infiltrate the PI foams' surface, in turn strongly bonding beads.

PEI foams have been easily prepared by solid-state foaming processes (temperature rise) after blowing agent solubilization in a high pressure vessel. Miller et al. investigated the solid-state batch foaming of polyetherimide (PEI) using subcritical $CO_2$ as a blowing agent (Miller, Chatchaisucha, and Kumar, 2009). They reported on the creation of microcellular and nanocellular morphologies in PEI foams, with 40% and higher relative density. A large solid-state foaming process window has been identified that allows for the creation of either microcellular or nanocellular structures at comparable density reductions (Figure 9.15). A transition from microscale

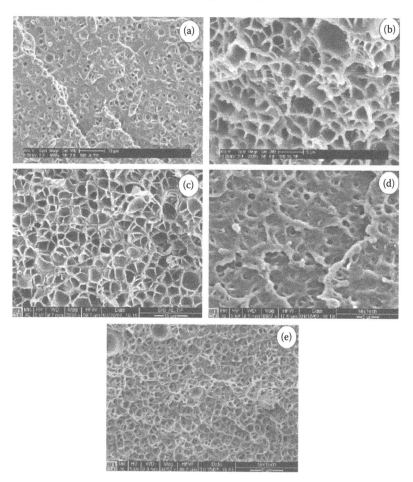

**FIGURE 9.15** Microcellular PEI structures created at 1 MPa (A–C) and 4 MPa (D,E) saturation pressures. (a) Sample bar = 10 μm, (b) Sample bar = 5 μm, (c) Sample bar = 10 μm, (d) Sample bar = 2 μm, (e) Sample bar = 5 μm. (Reported from Miller D. et al., 2009, *Polymer* 50, 23, November, 5576–5584, doi:10.1016/j.polymer.2009.09.020, http://linkinghub.elsevier.com/retrieve/pii/S0032386109007770.)

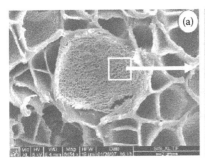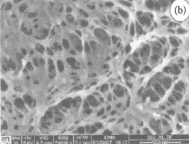

**FIGURE 9.16** Microcellular sample #9, 56.0% relative density, showing nano features on internal cell wall. (a) Bar = 2 μm, (b) Focus area of interior of cell wall shown in (a), (b) Bar = 500 nm. (Reported from Miller D. et al., 2009, *Polymer* 50, 23, November, 5576–5584, doi:10.1016/j.polymer.2009.09.020, http://linkinghub.elsevier.com/retrieve/pii/S0032386109007770.)

cells to nanoscale cells was observed at gas concentrations in the range of 94–110 mg $CO_2$/g PEI and a hierarchical structure was observed which consisted of nanocellular structures internal to microcells (Figure 9.16), as also experienced by Sorrentino, Aurilia, and Iannace (2011).

Nanocellular foam was prepared by the $CO_2$ temperature quench method from a high $T_g$ engineering polymers blend, namely PEEK with two types of PEI differing for the diamine in the structure (Nemoto, Takagi, and Ohshima, 2010; Pan, Zhan, and Wang, 2010). They prepared foams by templating the blend morphology of the immiscible PEEK/p-diamine PEI blends and foaming the amorphous PEI. The cell size was reduced to 100 nm and the bubble location was highly controlled in the p-PEI domain, aligned between the PEEK layers.

### 9.4.5 LIQUID CRYSTAL POLYMERS (LCP)

High-performance liquid crystalline polymers are polymers characterized by high melting temperatures (in the order of 280°C) and glass transition temperatures higher than 100°C. Their main property, from the processing point of view, is that they have very low viscosity above the melting temperature, and this renders them very compliant to injection-molding applications. This aspect, on the contrary, is counterproductive in the foaming process, for which high viscosity is needed. For this reason very poor investigations on their foamability have been performed, and they were only used in combination with amorphous polystyrene to prepare foams with extended service temperature range (Jin, 2001). In this case, foamed PS/LCP blends with cell diameter less than 7 μm have been produced by using supercritical $CO_2$, solubilized at 25 MPa, 80°C for 6 h. Due to poor adsorption of supercritical $CO_2$ by LCP under experimental conditions used by authors, the microvoids only exist in the polystyrene phase of the blends. A skin-core structure was detected in both unfoamed and foamed PS/LCP blends. In compositions where the compatibilizer, zinc sulfonated polystyrene ionomer (ZnSPS), was used, the microcellular blends exhibited larger cell size than those without ZnSPS and lower nucleated cells, as a consequence of the improved interfacial adhesion and the lack of heterogeneous sites for nucleating cells.

In order to improve the rheological properties of LCP polymers, the availability of nanocomposites could be a viable solution to prepare foams with good morphologies. MMT nanoparticles were used to prepare LCP nanocomposites by Bandyopadhyay et al. who experienced a significant increase in the rheological properties through the addition of 3.4% of dimethyl dehydrogenated tallow quaternary ammonium-modified MMT (Bandyopadhyay, Sinha Ray, and Bousmina, 2007, 2008).

Multi-wall carbon nanotubes (MWNTs) were also used as the reinforcing phase in LCP nanocomposites prepared by means of melt blending technique in a twin-screw extruder equipped with an ultrasonic unit to facilitate MWCNT dispersion (Kumar and Isayev, 2010). The role of ultrasonication was positive and resulted in increased structural as well as rheological properties because of the better-dispersed nanofiller. MWCNT were also used in LCP blends with polycarbonate (PC) (Mukherjee et al., 2009) and PEI (Nayak, Rajasekar, and Das, 2010). In the first case, PC/LCP/MWNTs nanocomposites containing as-received or modified (COOH-MWNT) carbon nanotubes were prepared through the melt process in an extruder and then compression molded. The incorporation of functionalized MWCNTs improved thermal, structural, dynamic-mechanical, and electrical properties of the composites, in particular in blends with treated MWCNTs. MWCNTs were also used, both unmodified and surface treated with SiC particles, to improve dispersion in PEI/LCP blends prepared by melt blending. In the ternary system, viscosity in the blend with modified MWCNTs was found to be lower than the ternary blend with pure MWCNTs, probably because modified MWCNTs improved the fibrillation of LCP compared to pure MWCNTs. Nanocomposite matrices have not been used to prepare foams yet.

### 9.4.6 Poly(Phenylene Sulfide) (PPS)

Poly(phenylene sulfide) is a semicrystalline polymer and, despite of its glass transition temperature of 85°C, is usually included in the high-performance polymers family thanks to its melting temperature equal to 285°C and its high degree of crystallinity, which gives it high elastic modulus and solvent resistance up to 110°C. No evidence of attempts to foam it are present in the literature, but some investigations have been made to reinforce it with nanoparticles.

Nanocomposite of $PPS/CaCO_3$ was prepared by a melt mixing process, which allowed a good dispersion of calcium carbonate nanoparticles when the filler content was below 5 wt%. $CaCO_3$ nanoparticles resulted in improvement of the nucleation of crystals, but retarded the mobility of polymer chains (Wang et al., 2006). Nanoparticles also resulted in a slight improvement in the tensile strength but significantly increased fracture toughness of 300%. Authors addressed this behavior to the fact that $CaCO_3$ nanoparticles could act as stress concentration sites, which can promote cavitation at the particles' boundaries during loading, leading to higher plastic deformation of the matrix and in turn improving fracture toughness.

Tungsten disulfide ($WS_2$) nanoparticles also were used to prepare PPS nanocomposites via melt blending (Naffakh et al., 2009). The addition of IF-$WS_2$

with concentrations greater than or equal to 0.5 wt% remarkably improved the mechanical performance of PPS with an increase in the storage modulus of 40–75% in dynamic mechanical tests.

The preparation and characterization of PPS nanocomposites based on polyetherimide-wrapped single-wall nanotubes and inorganic fullerene-like nanoparticles $WS_2$ has been reported by Díez-Pascual, Naffakh, Marco, and Ellis (2012), who used a melt blending technique to disperse nanoparticles. The wrapping of SWCNTs in polyetherimide (PEI) and the addition of inorganic fullerene-like tungsten disulfide (IF-WS$_2$) nanoparticles resulted in an effective method for improving the dispersion of SWCNTs, leading to enhanced properties. Mechanical tests showed significant enhancements in stiffness, strength, and toughness by the addition of both nanofillers. The electrical conductivity of PPS also increased dramatically at low SWCNT content (0.1–0.5 wt%), while the replacement of part of the SWCNTs with IF-WS2 did not result in a conductivity drop. Authors reported that the introduction of $WS_2$ nanoparticles in PPS/SWCNT nanocomposites gave higher performance than composites reinforced solely with wrapped or nonwrapped SWCNTs, and that properties were tailorable by modifying the SWCNT/IF-WS2 ratio.

### 9.4.7 SYNDIOTACTIC POLYSTYRENE

Syndiotactic polystyrene (sPS) is a semicrystalline polymer characterized by a glass transition temperature of 100°C and a melting temperature of 280°C. It has been widely investigated for the peculiar nanocavities that can be induced by carefully controlling its crystalline forms, but no studies are available at present on the foaming with blowing agents.

Some works on sPS are present, but essentially they regard the incorporation of organophilic clays (Park et al., 2001) and the characterization of sPS/clay nanocomposites with respect to the crystallization behavior (Tseng, Lee, and Chang, 2001; Wu et al., 2004), mechanical properties (Ho Kim et al., 2004), and moldability by means of injection-molding process (Sorrentino, Pantani, and Brucato, 2006). A work on the reinforcing of sPS by means of carbon nanocapsules is also present (Wang et al., 2008).

## REFERENCES

Ahn, S. H., S. H. Kim., and S. G. Lee. 2004. Surface-modified silica nanoparticle-reinforced poly(ethylene 2,6-naphthalate). *Journal of Applied Polymer Science* 94 (2) (October 15): 812–818. doi:10.1002/app.21007. http://doi.wiley.com/10.1002/app.21007.

Akartuna, I., E. Tervoort, J. C. H. Wong, A. R. Studart, and L. J. Gauckler. 2009. Macroporous polymers from particle-stabilized emulsions. *Polymer* 50 (15) (July): 3645–3651. doi:10.1016/j.polymer.2009.05.023. http://linkinghub.elsevier.com/retrieve/pii/S0032386109004297.

Alexandre, M. and P. Dubois. 2000. Polymer-layered silicate nanocomposites: Preparation, properties and uses of a new class of materials. *Materials Science and Engineering: R: Reports* 28 (1–2) (June): 1–63. doi:10.1016/S0927-796X(00)00012-7. http://linkinghub.elsevier.com/retrieve/pii/S0927796X00000127.

Al Ghatta, H. and S. Cobror. 1998. Polyester resins having improved rheological properties. U.S. Patent US5776994.

Arefmanesh, A. and S. G. Advani. 1991. Diffusion-induced growth of a gas bubble in a viscoelastic fluid. *Rheologica Acta* 30 (3): 274–283. doi:10.1007/s10019-002-0172-8. http://link.springer.com/10.1007/BF00366641.

Arefmanesh, A., S. G. Advani, and E. E. Michaelides. 1992. An accurate numerical solution for mass diffusion-induced bubble growth in viscous liquids containing limited dissolved gas. *International Journal of Heat and Mass Transfer* 35 (7) (July): 1711–1722. doi:10.1016/0017-9310(92)90141-E. http://linkinghub.elsevier.com/retrieve/pii/001793109290141E.

Aurilia, M., L. Sorrentino, and S. Iannace. 2012. Modelling physical properties of highly crystallized polyester reinforced with multiwalled carbon nanotubes. *European Polymer Journal* 48 (1) (January): 26–40. doi:10.1016/j.eurpolymj.2011.10.011. http://linkinghub.elsevier.com/retrieve/pii/S0014305711003909.

Aurilia, M., L. Sorrentino, L. Sanguigno, and S. Iannace. 2010. Nanofilled polyethersulfone as matrix for continuous glass fibers composites: Mechanical properties and solvent resistance. *Advances in Polymer Technology* 29 (3) (September 28): 146–160. doi:10.1002/adv.20187. http://doi.wiley.com/10.1002/adv.20187.

Balaji, V., A. N. Tiwari, and R. K. Goyal. 2011. Fabrication and properties of high performance PEEK/Si3N4 nanocomposites. *Journal of Applied Polymer Science* 119 (1) (January 5): 311–318. doi:10.1002/app.32750. http://doi.wiley.com/10.1002/app.32750.

Baldwin, D. F., C. B. Park, and N. P. Suh. 1996. A microcellular processing study of poly(ethylene terephthalate) in the amorphous and semicrystalline states. Part I: Microcell nucleation. *Polymer Engineering & Science* 36 (11) (May 15): 1437–1445. doi:10.1002/pen.10538. http://doi.wiley.com/10.1002/pen.10538.

Baldwin, D. F., M. Shimbo, and N. P. Suh. 1995. The role of gas dissolution and induced crystallization during microcellular polymer processing: A study of poly (ethylene terephthalate) and carbon dioxide systems. *Journal of Engineering Materials and Technology* 117 (1): 62. doi:10.1115/1.2804373. http://link.aip.org/link/JEMTA8/v117/i1/p62/s1&Agg=doi.

Bandyopadhyay, J., S. Sinha Ray, and M. Bousmina. 2007. Effect of organoclay on the orientation and thermal properties of liquid-crystalline. *Macromolecular Chemistry and Physics* 208 (18) (September 20): 1979–1991. doi:10.1002/macp.200700350. http://doi.wiley.com/10.1002/macp.200700350.

Bandyopadhyay, J., S. Sinha Ray, and M. Bousmina. 2008. Viscoelastic properties of clay-containing nanocomposites of thermotropic liquid-crystal polymer. *Macromolecular Chemistry and Physics* 210 (2) (December 17): 161–171. doi:10.1002/macp.200800479. http://doi.wiley.com/10.1002/macp.200800479.

Behrendt, N., C. Greiner, F. Fischer, T. Frese, V. Altstädt, H. W. Schmidt, R. Giesa, J. Hillenbrand, and G. M. Sessler. 2006. Morphology and electret behaviour of microcellular high glass temperature films. *Applied Physics* A 85 (1) (August 1): 87–93. doi:10.1007/s00339-006-3660-7. http://www.springerlink.com/index/10.1007/s00339-006-3660-7.

Bikiaris, D., V. Karavelidis, and G. Karayannidis. 2006. A new approach to prepare poly(ethylene terephthalate)/silica nanocomposites with increased molecular weight and fully adjustable branching or crosslinking by SSP. *Macromolecular Rapid Communications* 27 (15) (August 2): 1199–1205. doi:10.1002/marc.200600268. http://doi.wiley.com/10.1002/marc.200600268.

Blundell, D. J. and B. N. Osborn. 1983. The morphology of poly(aryl-ether-ether-ketone). *Polymer* 24 (8) (August): 953–958. doi:10.1016/0032-3861(83)90144-1. http://linkinghub.elsevier.com/retrieve/pii/0032386183901441.

Bratychak, M., W. Brostow, V. M. Castano, V. Donchak, and H. Gargai. 2002. Crosslinking agents of unsaturated polymers: Evaluation of the agent efficiency. *Materials Research Innovations* 6 (4): 153–159. doi:10.1007/s10019-002—0172-8.

Cai, H., F. Yan, and Q. Xue. 2004. Investigation of tribological properties of polyimide/carbon nanotube nanocomposites. *Materials Science and Engineering*: A 364 (1–2) (January): 94–100. doi:10.1016/S0921-5093(03)00669-5. http://linkinghub.elsevier.com/retrieve/pii/S0921509303006695.

Cai, H., F. Yan, Q. Xue, and W. Liu. 2003. Investigation of tribological properties of al2o3-polyimide nanocomposites. *Polymer Testing* 22 (8) (December): 875–882. doi:10.1016/S0142-9418(03)00024-2. http://linkinghub.elsevier.com/retrieve/pii/S0142941803000242.

Cano, C. I., E. S. Weiser, T. Kyu, and R. B. Pipes. 2005. Polyimide foams from powder: Experimental analysis of competitive diffusion phenomena. *Polymer* 46 (22) (October): 9296–9303. doi:10.1016/j.polymer.2005.07.056. http://linkinghub.elsevier.com/retrieve/pii/S0032386105010542.

Chae, H. G., B. C. Kim, S. S. Im, and Y. K. Han. 2001. Effect of molecular weight and branch structure on the crystallization and rheological properties of poly(butylene adipate). *Polymer Engineering & Science* 41 (7) (July): 1133–1139. doi:10.1002/pen.10814. http://doi.wiley.com/10.1002/pen.10814.

Chang, L., Z. Zhang, H. Zhang, and K. Friedrich. 2005. Effect of nanoparticles on the tribological behaviour of short carbon fibre reinforced poly(etherimide) composites. *Tribology International* 38 (11–12) (November): 966–973. doi:10.1016/j.triboint.2005.07.026. http://linkinghub.elsevier.com/retrieve/pii/S0301679X05001957.

Charlier, Y., J. L. Hedrick, T. P. Russell, A. Jonas, and W. Volksen. 1995. High temperature polymer nanofoams based on amorphous, high Tg polyimides. *Polymer* 36 (5) (March): 987–1002. doi:10.1016/0032-3861(95)93599-H. http://linkinghub.elsevier.com/retrieve/pii/003238619593599H.

Chen, Y. and J. O. Iroh. 1999. Synthesis and characterization of polyimide/silica hybrid composites. *Chemistry of Materials* 11 (5) (May): 1218–1222. doi:10.1021/cm980428l. http://pubs.acs.org/doi/abs/10.1021/cm980428l.

Choudhury, A. 2010. Dielectric and piezoelectric properties of polyetherimide/BaTiO3. *Nanocomposites. Materials Chemistry and Physics* 121 (1–2) (May): 280–285. doi:10.1016/j.matchemphys.2010.01.035. http://linkinghub.elsevier.com/retrieve/pii/S0254058410000593.

Colton, J. S. and N. P. Suh. 1986. The nucleation of microcellular thermoplastic foam: Process model and experimental results. *Advanced Manufacturing Processes* 1 (3-4) (January): 341–364. doi:10.1080/10426918608953169. http://www.tandfonline.com/doi/abs/10.1080/10426918608953169.

Colton, J. S. and N. P. Suh. 1987a. Nucleation of microcellular foam: Theory and practice. *Polymer Engineering and Science* 27 (7) (April): 500–503. doi:10.1002/pen.760270704. http://doi.wiley.com/10.1002/pen.760270704.

Colton, J. S. and N. P. Suh. 1987b. The nucleation of microcellular thermoplastic foam with additives: Part I: Theoretical considerations. *Polymer Engineering and Science* 27 (7) (April): 485–492. doi:10.1002/pen.760270702. http://doi.wiley.com/10.1002/pen.760270702.

Colton, J. S. and N. P. Suh. 1987c. The nucleation of microcellular thermoplastic foam with additives: Part II: Experimental results and discussion. *Polymer Engineering and Science* 27 (7) (April): 493–499. doi:10.1002/pen.760270703. http://doi.wiley.com/10.1002/pen.760270703.

Cotugno S., E. di Maio, G. Mensitieri, L. Nicolais, and S. Iannace. 2005. *Handbook of Biodegradable Polymeric Materials and Their Applications*. ed. S. K. Mallapragada and B. Narasimhan. Ames, IA: American Scientific Publishers.

Deng, F., T. Ogasawara, and N. Takeda. 2007. Evaluating the orientation and dispersion of carbon nanotubes inside nanocomposites by a focused-ion-beam technique. *Materials Letters* 61 (29) (December): 5095–5097. doi:10.1016/j.matlet.2007.04.049. http://linkinghub.elsevier.com/retrieve/pii/S0167577X0700376X.

Di, Y., S. Iannace, E. Di Maio, and L. Nicolais. 2005. Poly(lactic acid)/organoclay nano-composites: Thermal, rheological properties and foam processing. *Journal of Polymer Science Part B: Polymer Physics* 43 (6) (March 15): 689–698. doi:10.1002/polb.20366. http://doi.wiley.com/10.1002/polb.20366.

Di Maio, E. and S. Iannace. 2009. Foaming analysis of poly(epsilon-caprolactone) and poly(lactic acid) and their nanocomposites. In *Polymeric Foams: Technology and Developments in Regulations, Processes and Products*, ed. S. T. Lee and D. Scholtz. Boca Raton, FL: CRC Press/Taylor & Francis.

Díez-Pascual, A. M., G. Martínez, and M. A. Gómez. 2009. Synthesis and characterization of poly(ether ether ketone) derivatives obtained by carbonyl reduction. *Macromolecules* 42 (18) (September 22): 6885–6892. doi:10.1021/ma901208e. http://pubs.acs.org/doi/abs/10.1021/ma901208e.

Díez-Pascual, A. M., M. Naffakh, M. A. Gómez, C. Marco, G. Ellis, J. M. González-Domínguez, A Ansón, et al. 2009. The influence of a compatibilizer on the thermal and dynamic mechanical properties of PEEK/carbon nanotube composites. *Nanotechnology* 20 (31) (August 5): 315707. doi:10.1088/0957-4484/20/31/315707. http://www.ncbi.nlm.nih.gov/pubmed/19597256.

Díez-Pascual, A. M., M. Naffakh, J. M. González-Domínguez, A. Ansón, Y. Martínez-Rubi, M. T. Martínez, B. Simard, and M. A. Gómez. 2010. High performance PEEK/carbon nanotube composites compatibilized with polysulfones-I. Structure and thermal properties. *Carbon* 48 (12) (October): 3485–3499. doi:10.1016/j.carbon.2010.05.046. http://linkinghub.elsevier.com/retrieve/pii/S0008622310003830.

Díez-Pascual, A. M., M. Naffakh, C. Marco, and G. Ellis. 2012. Mechanical and electrical properties of carbon nanotube/poly(phenylene sulphide) composites incorporating poly-etherimide and inorganic fullerene-like nanoparticles. *Composites Part A: Applied Science and Manufacturing* 43 (4) (April): 603–612. doi:10.1016/j.compositesa.2011.12.026. http://linkinghub.elsevier.com/retrieve/pii/S1359835X12000036.

Díez-Pascual, A. M., M. Naffakh, C. Marco, G. Ellis, and M. A. Gómez-Fatou. 2012. High-performance nanocomposites based on polyetherketones. *Progress in Materials Science* 57 (7) (September): 1106–1190. doi:10.1016/j.pmatsci.2012.03.003. http://linkinghub.elsevier.com/retrieve/pii/S0079642512000266.

Durmus, A., N. Ercan, G. Soyubol, H. Deligöz, and A. Kaşgöz. 2009. Nonisothermal crystallization kinetics of poly(ethylene terephthalate)/clay nanocomposites prepared by melt processing. *Polymer Composites*: NA–NA. doi:10.1002/pc.20892. http://doi.wiley.com/10.1002/pc.20892.

Faghihi, K., M. Shabanian, and N. Bolhasani. 2012. Synthesis and preparation of new reinforced montmorillonite poly(amides-imides) based on n-trimellitimido-4-amino benzoic acid. *High Temperature Materials and Processes* 30 (6) (January). doi:10.1515/htmp.2011.113. http://www.degruyter.com/view/j/htmp.2011.30.issue-6/htmp.2011.113/htmp.2011.113.xml.

Fujimoto, Y., S. Sinha Ray, M. Okamoto, A. Ogami, K. Yamada, and K. Ueda. 2003. Well-controlled biodegradable nanocomposite foams: From microcellular to nanocellular. *Macromolecular Rapid Communications* 24 (7) (May 7): 457–461. doi:10.1002/marc.200390068. http://doi.wiley.com/10.1002/marc.200390068.

Ghaemi, N., S. S. Madaeni, A. Alizadeh, H. Rajabi, and P. Daraei. 2011. Preparation, characterization and performance of polyethersulfone/organically modified montmo-rillonite nanocomposite membranes in removal of pesticides. *Journal of Membrane Science* 382 (1–2) (October): 135–147. doi:10.1016/j.memsci.2011.08.004. http://linkinghub.elsevier.com/retrieve/pii/S0376738811005813.

Ghasemi, H., P. J. Carreau, and M. R. Kamal. 2012. Isothermal and non-isothermal crystallization behavior of PET nanocomposites. *Polymer Engineering & Science* 52 (2) (February 8): 372–384. doi:10.1002/pen.22092. http://doi.wiley.com/10.1002/pen.22092.

Ghasemi, H., P. J. Carreau, M. R. Kamal, and J. Uribe-Calderon. 2011. Preparation and characterization of PET/clay nanocomposites by melt compounding. *Polymer Engineering & Science* 51 (6) (June 11): 1178–1187. doi:10.1002/pen.21874. http://doi. wiley.com/10.1002/pen.21874.

Giannelis, E. P., R. Krishnamoorti, and E. Manias. 1999. Polymer-silicate nanocomposites: Model systems for confined polymers and polymer brushes. *Advances in Polymer Science* 138: 107–147. doi:10.1007/3-540-69711-X_3. http://link.springer.com/ chapter/10.1007/3-540-69711-X_3?LI=true.

Gibson, L. and M. Ashby. 1997. *Cellular Solids: Structure and Properties*, 2nd ed. Cambridge: Cambridge University Press.

Goel, S. K. and E. J. Beckman. 1993. Plasticization of poly(methyl methacrylate) (PMMA) networks by supercritical carbon dioxide. *Polymer* 34 (7) (January): 1410–1417. doi:10.1016/0032-3861(93)90853-3. http://linkinghub.elsevier.com/retrieve/pii/0032386193908533.

Goel, S. K. and E. J. Beckman. 1995. Nucleation and growth in microcellular materials: Supercritical CO2 as foaming agent. *AIChE Journal* 41 (2) (February): 357–367. doi:10.1002/aic.690410217. http://doi.wiley.com/10.1002/aic.690410217.

Golubeva, O. Y., V. E. Yudin, A. L. Didenko, V. M. Svetlichnyi, and V. V. Gusarov. 2007. Nanocomposites based on polyimide thermoplastics and magnesium silicate nanoparticles with montmorillonite structure. *Russian Journal of Applied Chemistry* 80 (1) (January): 106–109. doi:10.1134/S1070427207010211. http://www.springerlink.com/index/10.1134/S1070427207010211.

Goyal, R. K. 2009. Thermal, mechanical, and dielectric properties of high performance PEEK/AlN nanocomposites. *Journal of Nanoscience and Nanotechnology* 9 (12). doi:10.1166/jnn.2009.1582. http://www.ingentaconnect.com/content/asp/jnn/2009/00000009/00000012/art00016.

Guo, Q., J. Wang, C. B Park, and M. Ohshima. 2006. A microcellular foaming simulation system with a high pressure-drop rate. *Industrial & Engineering Chemistry Research* 45 (18) (August): 6153–6161. doi:10.1021/ie060105w. http://pubs.acs.org/doi/abs/10.1021/ie060105w.

Handa, Y. P. and Z. Zhang. 2000. A new technique for measuring retrograde vitrification in polymer-gas systems and for making ultramicrocellular foams from the retrograde phase. *Journal of Polymer Science Part B: Polymer Physics* 38 (5) (March 1): 716–725. doi:10.1002/(SICI)1099-0488(20000301)38:5<716::AID-POLB9>3.0.CO;2-N. http://doi.wiley.com/10.1002/%28SICI%291099-0488%2820000301%2938%3A5%3C716%3A%3AAID-POLB9%3E3.0.CO%3B2-N.

Harris, L. D., B. S. Kim, and D. J. Mooney. 1998. Open pore biodegradable matrices formed with gas foaming. *Journal of Biomedical Materials Research* 42 (3): 396–402. doi:10.1002/(SICI)1097-4636(19981205)42:3<396::AID-JBM7>3.0.CO;2-E.

Hayrapetyan, S., A. Kelarakis, L. Estevez, Q. Lin, K. Dana, Y. L. Chung, and E. P. Giannelis. 2012. Non-toxic poly(ethylene terephthalate)/clay nanocomposites with enhanced barrier properties. *Polymer* 53 (2) (January): 422–426. doi:10.1016/j.polymer.2011.12.017. http://linkinghub.elsevier.com/retrieve/pii/S0032386111010238.

Hedrick, J. 1995. High temperature nanofoams derived from rigid and semi-rigid polyimides. *Polymer* 36 (14): 2685–2697. doi:10.1016/0032-3861(95)93645-3. http://linkinghub.elsevier.com/retrieve/pii/0032386195936453.

Hedrick, J. L., K. R. Carter, R. Richter, R. D. Miller, T. P. Russell, V. Flores, D. Meccereyes, P. Dubois, and R. Jérôme. 1998. Polyimide nanofoams from aliphatic polyester-based copolymers. *Chemistry of Materials* 10 (1) (January): 39–49. doi:10.1021/cm960523z. http://pubs.acs.org/doi/abs/10.1021/cm960523z.

Hedrick, J. L., R. DiPietro, C. J. G. Plummer, J. Hilborn, and R. Jerome. 1996. Polyimide foams derived from a high Tg polyimide with grafted poly(α-methylstyrene). *Polymer* 37 (23) (November): 5229–5236. doi:10.1016/0032-3861(96)00331-X. http://linkinghub.elsevier.com/retrieve/pii/003238619600331X.

Ho Kim M., C. I. Park, W. M. Choi, J. W. Lee, J. G. Lim, O. O. Park, and J. M. Kim. 2004. Synthesis and material properties of syndiotactic polystyrene/organophilic clay nanocomposites. *Journal of Applied Polymer Science* 92 (4) (May 15): 2144–2150. doi:10.1002/app.20186. http://doi.wiley.com/10.1002/app.20186.

Huang, C. C., G. W. Jang, K. C. Chang, W. I Hung, and J. M. Yeh. 2008. High-performance polyimide-clay nanocomposite materials based on a dual intercalating agent system. *Polymer International* 57 (4) (April): 605–611. doi:10.1002/pi.2381. http://doi.wiley.com/10.1002/pi.2381.

Huang, J., K. Zhang, K. Wang, Z. Xie, B. Ladewig, and H. Wang. 2012. Fabrication of polyethersulfone-mesoporous silica nanocomposite ultrafiltration membranes with antifouling properties. *Journal of Membrane Science* 423–424 (December): 362–370. doi:10.1016/j.memsci.2012.08.029. http://linkinghub.elsevier.com/retrieve/pii/S037673881200628X.

Huang, J. C., Z. K. Zhu, X. F. Qian, and Y. Y. Sun. 2001. Poly(etherimide)/montmorillonite nanocomposites prepared by melt intercalation: Morphology, solvent resistance properties and thermal properties. *Polymer* 42 (3) (February): 873–877. doi:10.1016/S0032-3861(00)00411-0. http://linkinghub.elsevier.com/retrieve/pii/S0032386100004110.

Isayev, A. I., R. Kumar, and T. M. Lewis. 2009. Ultrasound assisted twin screw extrusion of polymer–nanocomposites containing carbon nanotubes. *Polymer* 50 (1) (January): 250–260. doi:10.1016/j.polymer.2008.10.052. http://linkinghub.elsevier.com/retrieve/pii/S0032386108009488.

Jaafar, J., A. F. Ismail, and T. Matsuura. 2012. Effect of dispersion state of Cloisite15A® on the performance of SPEEK/Cloisite15A nanocomposite membrane for DMFC application. *Journal of Applied Polymer Science* 124 (2) (April 15): 969–977. doi:10.1002/app.35139. http://doi.wiley.com/10.1002/app.35139.

Japon, S., L. Boogh, Y. Leterrier, and J. A. E. Månson. 2000. Reactive processing of poly(ethylene terephthalate) modified with multifunctional epoxy-based additives. *Polymer* 41 (15) (July): 5809–5818. doi:10.1016/S0032-3861(99)00768-5. http://linkinghub.elsevier.com/retrieve/pii/S0032386199007685.

Jin, W. 2001. An investigation on the microcellular structure of polystyrene/LCP blends prepared by using supercritical carbon dioxide. *Polymer* 42 (19) (September): 8265–8275. doi:10.1016/S0032-3861(01)00343-3. http://linkinghub.elsevier.com/retrieve/pii/S0032386101003433.

Ke, Y., C. Long, and Z. Qi. 1999. Crystallization, properties, and crystal and nanoscale morphology of PET-clay nanocomposites. *Journal of Applied Polymer Science* 71 (7) (February 14): 1139–1146. doi:10.1002/(SICI)1097-4628(19990214)71:7<1139::AID-APP12>3.0.CO;2-E. http://doi.wiley.com/10.1002/(SICI)1097-4628(19990214)71:7<1139::AID-APP12>3.0.CO;2-E.

Kim, J. Y., H. J. Choi, C. S. Kang, and S. H. Kim. 2009. Influence of modified carbon nanotube on physical properties and crystallization behavior of poly(ethylene terephthalate) nanocomposite. *Polymer Composites*: NA–NA. doi:10.1002/pc.20868. http://doi.wiley.com/10.1002/pc.20868.

Kim, J. Y., S. I. Han, and S. H. Kim. 2007. Crystallization behaviors and mechanical properties of poly(ethylene 2, 6-naphthalate)/multiwall carbon nanotube nanocomposites. *Polymer Engineering & Science* 47 (11) (November): 1715–1723. doi:10.1002/pen.20789. http://doi.wiley.com/10.1002/pen.20789.

Kim, J. Y., S. I. Han, D. K. Kim, and S. H. Kim. 2009. Mechanical reinforcement and crystallization behavior of poly(ethylene 2, 6-naphthalate) nanocomposites induced by modified carbon nanotube. *Composites Part A: Applied Science and Manufacturing* 40 (1) (January): 45–53. doi:10.1016/j.compositesa.2008.10.002. http://linkinghub.elsevier.com/retrieve/pii/S1359835X08002601.

Kim, J. Y. and S. H. Kim. 2006. Influence of multiwall carbon nanotube on physical properties of poly(ethylene 2,6-naphthalate) nanocomposites. *Journal of Polymer Science Part B: Polymer Physics* 44 (7) (April 1): 1062–1071. doi:10.1002/polb.20728. http://doi.wiley.com/10.1002/polb.20728.

Kim, J. Y., D. K. Kim, and S. H. Kim. 2009. Thermal decomposition behavior of poly(ethylene 2, 6-naphthalate)/silica nanocomposites. *Polymer Composites* 30 (12) (December): 1779–1787. doi:10.1002/pc.20749. http://doi.wiley.com/10.1002/pc.20749.

Kim, H. and C. W. Macosko. 2008. Morphology and properties of polyester/exfoliated graphite nanocomposites. *Macromolecules* 41 (9) (May): 3317–3327. doi:10.1021/ma702385h. http://pubs.acs.org/doi/abs/10.1021/ma702385h.

Klempner, C. and K. C. Frish, eds. 1991. *Handbook of Polymeric Foams and Foam Technology.* New York: Hanser.

Klempner, D. and V. Sendjarevic. 2000. *Polymeric Foams and Foam Technology.* 2nd. ed. New York: Hanser Publishers.

Knör, N., A. Gebhard, F. Haupert, and A. K. Schlarb. 2009. Polyetheretherketone (PEEK) nanocomposites for extreme mechanical and tribological loads. *Mechanics of Composite Materials* 45 (2) (May 27): 199–206. doi:10.1007/s11029-009-9071-z. http://www.springerlink.com/index/10.1007/s11029-009-9071-z.

Krause, B., K. Diekmann, N. F. A. van der Vegt, and M. Wessling. 2002. Open nanoporous morphologies from polymeric blends by carbon dioxide foaming. *Macromolecules* 35 (5) (February): 1738–1745. doi:10.1021/ma011672s. http://pubs.acs.org/doi/abs/10.1021/ma011672s.

Krause, B., R. Mettinkhof, N. F. A. van der Vegt, and M. Wessling. 2001. Microcellular foaming of amorphous high-Tg polymers using carbon dioxide. *Macromolecules* 34 (4) (February): 874–884. doi:10.1021/ma001291z. http://pubs.acs.org/doi/abs/10.1021/ma001291z.

Krause, B., H. J. P. Sijbesma, P. Münüklü, N. F. A. van der Vegt, and M. Wessling. 2001. Bicontinuous nanoporous polymers by carbon dioxide foaming. *Macromolecules* 34 (25) (December): 8792–8801. doi:10.1021/ma010854j. http://pubs.acs.org/doi/abs/10.1021/ma010854j.

Kumar, R. and A. I. Isayev. 2010. Thermotropic LCP/CNF nanocomposites prepared with aid of ultrasonic waves. *Polymer* 51 (15) (July 8): 3503–3511. doi:10.1016/j.polymer.2010.05.042. http://linkinghub.elsevier.com/retrieve/pii/S0032386110004702.

Kumar, V. and J. Weller. 1994. Production of microcellular polycarbonate using carbon dioxide for bubble nucleation. *Journal of Engineering for Industry* 116 (4): 413. doi:10.1115/1.2902122. http://link.aip.org/link/JEFIA8/v116/i4/p413/s1&Agg=doi.

Kuo, M. C., J. S. Kuo, M. H. Yang, and J. C. Huang. 2010. On the crystallization behavior of the nano-silica filled PEEK composites. *Materials Chemistry and Physics* 123 (2–3) (October): 471–480. doi:10.1016/j.matchemphys.2010.04.043. http://linkinghub.elsevier.com/retrieve/pii/S0254058410003524.

LeBaron, P. 1999. Polymer-layered silicate nanocomposites: An overview. *Applied Clay Science* 15 (1–2) (September): 11–29. doi:10.1016/S0169-1317(99)00017-4. http://linkinghub.elsevier.com/retrieve/pii/S0169131799000174.

Lee, S. T., ed. 2000. *Foam Extrusion.* Lancaster: Technomic Publishing Co. Inc.

Lee, S. H., S. H. Choi, S. Y. Kim, and J. R. Youn. 2010. Effects of thermal imidization on mechanical properties of poly(amide-co-imide)/multiwalled carbon nanotube composite films. *Journal of Applied Polymer Science*: n/a–n/a. doi:10.1002/app.32177. http://doi.wiley.com/10.1002/app.32177.

Lee, L., C. Zeng, X. Cao, X. Han, J. Shen, and G. Xu. 2005. Polymer nanocomposite foams. *Composites Science and Technology* 65 (15–16) (December): 2344–2363. doi:10.1016/j.compscitech.2005.06.016. http://linkinghub.elsevier.com/retrieve/pii/S0266353805002253.

Li, M. and Y. G. Jeong. 2011. Poly(ethylene terephthalate)/exfoliated graphite nanocomposites with improved thermal stability, mechanical and electrical properties. *Composites Part A: Applied Science and Manufacturing* 42 (5) (May): 560–566. doi:10.1016/j.compositesa.2011.01.015. http://linkinghub.elsevier.com/retrieve/pii/S1359835X11000388.

Li, J. F., Z. L. Xu, H. Yang, L. Y. Yu, and M. Liu. 2009. Effect of TiO2 nanoparticles on the surface morphology and performance of microporous PES membrane. *Applied Surface Science* 255 (9) (February): 4725–4732. doi:10.1016/j.apsusc.2008.07.139. http://linkinghub.elsevier.com/retrieve/pii/S0169433208017674.

Liang, Z. 2003. Polyimide/montmorillonite nanocomposites based on thermally stable, rigid-rod aromatic amine modifiers. *Polymer* 44 (5) (March): 1391–1399. doi:10.1016/S0032-3861(02)00911-4. http://linkinghub.elsevier.com/retrieve/pii/S0032386102009114.

Liang, C. Y., P. Uchytil, R. Petrychkovych, Y. C. Lai, K. Friess, M. Sipek, M. M. Reddy, and S. Y. Suen. 2012. A comparison on gas separation between PES (polyethersulfone)/MMT (Na-montmorillonite) and PES/TiO2 mixed matrix membranes. *Separation and Purification Technology* 92 (May): 57–63. doi:10.1016/j.seppur.2012.03.016. http://linkinghub.elsevier.com/retrieve/pii/S1383586612001712.

Liang, Z. M. and J. Yin. 2003. Poly(etherimide)/montmorillonite nanocomposites prepared by melt intercalation. *Journal of Applied Polymer Science* 90 (7) (November 14): 1857–1863. doi:10.1002/app.12847. http://doi.wiley.com/10.1002/app.12847.

Liu, W., X. Tian, P. Cui, Y. Li, K. Zheng, and Y. Yang. 2004. Preparation and characterization of PET/silica nanocomposites. *Journal of Applied Polymer Science* 91 (2) (January 15): 1229–1232. doi:10.1002/app.13284. http://doi.wiley.com/10.1002/app.13284.

Liu, T., Y. Tong, and W. D. Zhang. 2007. Preparation and characterization of carbon nanotube/polyetherimide nanocomposite films. *Composites Science and Technology* 67 (3–4) (March): 406–412. doi:10.1016/j.compscitech.2006.09.007. http://linkinghub.elsevier.com/retrieve/pii/S0266353806003526.

Madaeni, S. S., M. Moahamadi, S. Badieh, V. Vatanpour, and N. Ghaemi. 2012. Effect of titanium dioxide nanoparticles on polydimethylsiloxane/polyethersulfone composite membranes for gas separation. *Polymer Engineering & Science* 52 (12) (December 19): 2664–2674. doi:10.1002/pen.23223. http://doi.wiley.com/10.1002/pen.23223.

Marrazzo, C., E. Di Maio, and S. Iannace. 2008. Conventional and nanometric nucleating agents in poly(ε-caprolactone) foaming: Crystals versus bubbles nucleation. *Polymer Engineering & Science* 48 (2) (February): 336–344. doi:10.1002/pen.20937. http://doi.wiley.com/10.1002/pen.20937.

May-Pat, A. 2011. Mechanical properties of pet composites using multi-walled carbon nanotubes functionalized by inorganic and itaconic acids. *Express Polymer Letters* 6 (2) (December 1): 96–106. doi:10.3144/expresspolymlett.2012.11. http://www.expresspolymlett.com/letolt.php?file = EPL-0002773&mi = c.

Miller, D., P. Chatchaisucha, and V. Kumar. 2009. Microcellular and nanocellular solid-state polyetherimide (pei) foams using sub-critical carbon dioxide I. Processing and structure. *Polymer* 50 (23) (November): 5576–5584. doi:10.1016/j.polymer.2009.09.020. http://linkinghub.elsevier.com/retrieve/pii/S0032386109007770.

Mishra, T. K., A. Kumar, V. Verma, K. N. Pandey, and V. Kumar. 2012. PEEK composites reinforced with zirconia nanofiller. *Composites Science and Technology* 72 (13) (August): 1627–1631. doi:10.1016/j.compscitech.2012.06.019. http://linkinghub.elsevier.com/retrieve/pii/S0266353812002436.

Mitsunaga, M., Y. Ito, S. S. Ray, M. Okamoto, and K. Hironaka. 2003. Intercalated polycarbonate/clay nanocomposites: Nanostructure control and foam processing. *Macromolecular Materials and Engineering* 288 (7) (July): 543–548. doi:10.1002/mame.200300097. http://doi.wiley.com/10.1002/mame.200300097.

Moniruzzaman, M. and K. I. Winey. 2006. Polymer nanocomposites containing carbon nanotubes. *Macromolecules* 39 (16) (August): 5194–5205. doi:10.1021/ma060733p. http://pubs.acs.org/doi/abs/10.1021/ma060733p.

Mooney, D. J., D. F. Baldwin, N. P. Suh, J. P. Vacanti, and R. Langer. 1996. Novel approach to fabricate porous sponges of poly(d,l-lactic-co-glycolic acid) without the use of organic solvents. *Biomaterials* 17 (14) (July): 1417–1422. doi:10.1016/0142-9612(96)87284-X. http://linkinghub.elsevier.com/retrieve/pii/014296129687284X.

Mukherjee, M., T. Das, R. Rajasekar, S. Bose, S. Kumar, and C. K. Das. 2009. Improvement of the properties of PC/LCP blends in the presence of carbon nanotubes. *Composites Part A: Applied Science and Manufacturing* 40 (8) (August): 1291–1298. doi:10.1016/j.compositesa.2009.05.024. http://linkinghub.elsevier.com/retrieve/pii/S1359835X09001742.

Naffakh, M., A. M. Díez-Pascual, C. Marco, M. A. Gómez, and I. Jiménez. 2010. Novel melt-processable poly(ether ether ketone)(peek)/inorganic fullerene-like WS(2) nanoparticles for critical applications. *The Journal of Physical Chemistry B* 114 (35) (September 9): 11444–53. doi:10.1021/jp105340g. http://www.ncbi.nlm.nih.gov/pubmed/20722359.

Naffakh, M., C. Marco, M. A Gómez, J. Gómez-Herrero, and I. Jiménez. 2009. Use of inorganic fullerene-like WS2 to produce new high-performance polyphenylene sulfide nanocomposites: Role of the nanoparticle concentration. *The Journal of Physical Chemistry B* 113 (30) (July 30): 10104–11. doi:10.1021/jp902700x. http://www.ncbi.nlm.nih.gov/pubmed/19719278.

Nayak, G. C., R. Rajasekar, and C. K. Das. 2010. Effect of SiC coated MWCNTs on the thermal and mechanical properties of PEI/LCP blend. *Composites Part A: Applied Science and Manufacturing* 41 (11) (November): 1662–1667. doi:10.1016/j.compositesa.2010.08.003. http://linkinghub.elsevier.com/retrieve/pii/S1359835X10002150.

Nemoto, T., J. Takagi, and M. Ohshima. 2010. Nanocellular foams-cell structure difference between immiscible and miscible PEEK/PEI polymer blends. *Polymer Engineering & Science* 50 (12) (December 30): 2408–2416. doi:10.1002/pen.21766. http://onlinelibrary.wiley.com/doi/10.1002/pen.21766/full.

Nguyen, Q. T., S. Japon, A. Luciani, Y. Leterrier, and J. A. E. Månson. 2001. Molecular characterization and rheological properties of modified poly(ethylene terephthalate) obtained by reactive extrusion. *Polymer Engineering & Science* 41 (8) (August): 1299–1309. doi:10.1002/pen.10830. http://doi.wiley.com/10.1002/pen.10830.

Oh, S. J., H. J. Lee, D. K. Keum, S. W. Lee, D. H. Wang, S. Y. Park, L. S. Tan, and J. B. Baek. 2006. Multiwalled carbon nanotubes and nanofibers grafted with polyetherketones in mild and viscous polymeric acid. *Polymer* 47 (4) (February): 1132–1140. doi:10.1016/j.polymer.2005.12.064. http://linkinghub.elsevier.com/retrieve/pii/S0032386105018239.

Özen, İ., F. İnceoğlu, K. Acatay, and Y. Z. Menceloğlu. 2012. Comparison of melt extrusion and thermokinetic mixing methods in poly(ethylene terephthalate)/montmorillonite nanocomposites. *Polymer Engineering & Science* 52 (7) (July 21): 1537–1547. doi:10.1002/pen.23102. http://doi.wiley.com/10.1002/pen.23102.

Pan, L. Y., M. S. Zhan, and K. Wang. 2010. Preparation and characterization of high-temperature resistance polyimide foams. *Polymer Engineering & Science* 50 (6) (June 18): 1261–1267. doi:10.1002/pen.21653. http://doi.wiley.com/10.1002/pen.21653.

Park, J., W. Choi, J. Cho, B. H. Chun, S. H. Kim, K. B. Lee, and J. Bang. 2010. Carbon nanotube-based nanocomposite desalination membranes from layer-by-layer assembly. *Desalination and Water Treatment* 15 (1–3) (March): 76–83. doi:10.5004/dwt.2010.1670. http://www.tandfonline.com/doi/abs/10.5004/dwt.2010.1670.

Park, C. I., O. O. Park, J. G. Lim, and H. J. Kim. 2001. The fabrication of syndiotactic polystyrene/organophilic clay nanocomposites and their properties. *Polymer* 42 (17) (August): 7465–7475. doi:10.1016/S0032-3861(01)00213-0. http://linkinghub.elsevier.com/retrieve/pii/S0032386101002130.

Paterson, R., Y. Yampol'skii, P. G. T. Fogg, A. Bokarev, V. Bondar, O. Ilinich, and S. Shishatskii. 1999. IUPAC-NIST Solubility data series 70. Solubility of gases in glassy polymers. *Journal of Physical and Chemical Reference Data* 28 (5): 1255. doi:10.1063/1.556050. http://link.aip.org/link/JPCRBU/v28/i5/p1255/s1&Agg = doi.

Paul H. Y., B. Wong, Z. Zhang, V. Kumar, S. Eddy, and K. Khemani. 1999. Some thermodynamic and kinetic properties of the system PETG-CO2, and morphological characteristics of the CO2-blown PETG foams. *Polymer Engineering & Science* 39 (1) (January): 55–61. doi:10.1002/pen.11396. http://doi.wiley.com/10.1002/pen.11396.

Pavel, D. and R. Shanks. 2003. Molecular dynamics simulation of diffusion of O2 and CO2 in amorphous poly(ethylene terephthalate) and related aromatic polyesters. *Polymer* 44 (21) (October): 6713–6724. doi:10.1016/j.polymer.2003.08.016. http://linkinghub.elsevier.com/retrieve/pii/S003238610300764X.

Pontiff, T. 2000. Foaming agents for foam extrusion. In *Foam Extrusion*, ed. S. T. Lee. Lancaster: Technomic Publishing Company, Inc.

Ramesh, N. S., D. H. Rasmussen, and G. A. Campbell. 1991. Numerical and experimental studies of bubble growth during the microcellular foaming process. *Polymer Engineering and Science* 31 (23) (December): 1657–1664. doi:10.1002/pen.760312305. http://doi.wiley.com/10.1002/pen.760312305.

Ranade, A., N. A. D'Souza, and B. Gnade. 2002. Exfoliated and intercalated polyamide-imide nanocomposites with montmorillonite. *Polymer* 43 (13) (June): 3759–3766. doi:10.1016/S0032-3861(02)00106-4. http://linkinghub.elsevier.com/retrieve/pii/S0032386102001064.

Rhein, R. A. and J. D. Ingham. 1973. New polymer systems: Chain extension by dianhydrides. *Polymer* 14 (10) (October): 466–468. doi:10.1016/0032-3861(73)90151-1. http://linkinghub.elsevier.com/retrieve/pii/0032386173901511.

Sandler, J., A. H. Windle, P. Werner, V. Altst Adt, and M. V. Es. 2003. Carbon-nanofibre-reinforced poly(ether ether ketone). *Fibres* 38 (10): 2135–2141.

Scamardella, A. M., U. Vietri, L. Sorrentino, M. Lavorgna, and E. Amendola. 2012. Foams based on poly(ethylene terephthalate) nanocomposites with enhanced thermal stability. *Journal of Cellular Plastics* 48 (6) (November 20): 557–576. doi:10.1177/0021955X12445405. http://cel.sagepub.com/cgi/doi/10.1177/0021955X12445405.

Shantalii, T. A., I. L. Karpova, K. S. Dragan, E. G. Privalko, V. M. Karaman, and V. P. Privalko. 2005. Properties of an organosilicon nanophase generated in a poly(amide imide) matrix by the sol-gel technique. *Polymers for Advanced Technologies* 16 (5) (May): 400–404. doi:10.1002/pat.596. http://doi.wiley.com/10.1002/pat.596.

Shao, L., Y. P. Bai, X. Huang, L. H. Meng, and J. Ma. 2009. Fabrication and characterization of solution cast MWNTs/PEI nanocomposites. *Journal of Applied Polymer Science* 113 (3) (August 5): 1879–1886. doi:10.1002/app.30197. http://doi.wiley.com/10.1002/app.30197.

Shen, J., C. Zeng, and L. J. Lee. 2005. Synthesis of polystyrene–carbon nanofibers nanocomposite foams. *Polymer* 46 (14) (June): 5218–5224. doi:10.1016/j.polymer.2005.04.010. http://linkinghub.elsevier.com/retrieve/pii/S0032386105004349.

Shepard, K. B., H. Gevgilili, M. Ocampo, J. Li, F. T. Fisher, and D. M. Kalyon. 2012. Viscoelastic behavior of poly(ether imide) incorporated with multiwalled carbon nanotubes. *Journal of Polymer Science Part B: Polymer Physics* 50 (21) (November 1): 1504–1514. doi:10.1002/polb.23151. http://doi.wiley.com/10.1002/polb.23151.

Siochi, E. J., D. C. Working, C. Park, P. T. Lillehei, J. H. Rouse, C. C. Topping, A. R. Bhattacharyya, and S. Kumar. 2004. Melt processing of SWCNT-polyimide nanocomposite fibers. *Composites Part B: Engineering* 35 (5) (July): 439–446. doi:10.1016/j.compositesb.2003.09.007. http://linkinghub.elsevier.com/retrieve/pii/S1359836804000502.

Son, M., Y. Ha, M. C. Choi, T. Lee, D. Han, S. Han, and C. S. Ha. 2008. Microstructure and properties of polyamideimide/silica hybrids compatibilized with 3-aminopropyltriethoxysilane. *European Polymer Journal* 44 (7) (July): 2236–2243. doi:10.1016/j.eurpolymj.2008.04.037. http://linkinghub.elsevier.com/retrieve/pii/S0014305708002061.

Sorrentino, L., M. Aurilia, L. Cafiero, S. Cioffi, and S. Iannace. 2012. Mechanical behavior of solid and foamed polyester/expanded graphite nanocomposites. *Journal of Cellular Plastics* 48 (4) (July 12): 355–368. doi:10.1177/0021955X12449641. http://cel.sagepub.com/cgi/content/abstract/48/4/355.

Sorrentino, L., M. Aurilia, L. Cafiero, and S. Iannace. 2011. Nanocomposite foams from high-performance thermoplastics. *Journal of Applied Polymer Science* 122 (6) (December 15): 3701–3710. doi:10.1002/app.34784. http://doi.wiley.com/10.1002/app.34784.

Sorrentino, L., M. Aurilia, and S. Iannace. 2011. Polymeric foams from high-performance thermoplastics. *Advances in Polymer Technology* 30 (3) (September 8): 234–243. doi:10.1002/adv.20219. http://onlinelibrary.wiley.com/doi/10.1002/adv.20219/full.

Sorrentino, L, E. Di Maio, and S. Iannace. 2010. Poly(ethylene terephthalate) foams: Correlation between the polymer properties and the foaming process. *Journal of Applied Polymer Science* 116 (1) (April 5): 27–35. doi:10.1002/app.31427. http://onlinelibrary.wiley.com/doi/10.1002/app.31427/full.

Sorrentino, L., S. Iannace, E. Di Maio, and D. Acierno. 2005. Isothermal crystallization kinetics of chain-extended PET. *Journal of Polymer Science Part B: Polymer Physics* 43 (15) (August 1): 1966–1972. doi:10.1002/polb.20480. http://doi.wiley.com/10.1002/polb.20480.

Sorrentino, A., R. Pantani, and V. Brucato. 2006. Injection molding of syndiotactic polystyrene/clay nanocomposites. *Polymer Engineering & Science* 46 (12) (December): 1768–1777. doi:10.1002/pen.20650. http://doi.wiley.com/10.1002/pen.20650.

Sun, H. and J. E. Mark. 2002. Preparation, characterization, and mechanical properties of some microcellular polysulfone foams. *Journal of Applied Polymer Science* 86 (7) (November 14): 1692–1701. doi:10.1002/app.11070. http://doi.wiley.com/10.1002/app.11070.

Sun, H., G. S. Sur, and J. E Mark. 2002. Microcellular foams from polyethersulfone and poly-phenylsulfone. *European Polymer Journal* 38 (12) (December): 2373–2381. doi:10.1016/S0014-3057(02)00149-0. http://linkinghub.elsevier.com/retrieve/pii/S0014305702001490.

Tan, S. C., Z. Bai, H. Sun, J. E. Mark, F. E. Arnold, and C. Y. C. Lee. 2003. Processing of micro-cellular foams from polybenzobisthiazole/polyetherketone ketone molecular composites. *Journal of Materials Science* 38 (19): 4013–4019. doi:10.1023/A:1026218817102.

Tang, M., Y. C. Huang, and Y. P. Chen. 2004. Sorption and diffusion of supercritical carbon dioxide into polysulfone. *Journal of Applied Polymer Science* 94 (2) (October 15): 474–482. doi:10.1002/app.20895. http://doi.wiley.com/10.1002/app.20895.

Thostenson, E. T., Z. Ren, and T. W. Chou. 2001. Advances in the science and technology of carbon nanotubes and their composites: A review. *Composites Science and Technology* 61 (13) (October): 1899–1912. doi:10.1016/S0266-3538(01)00094-X. http://linkinghub.elsevier.com/retrieve/pii/S026635380100094X.

Tsai, T. Y., C. H. Li, C. H. Chang, W. H. Cheng, C. L. Hwang, and R. J. Wu. 2005. Preparation of exfoliated polyester/clay nanocomposites. *Advanced Materials* 17 (14) (July 18): 1769–1773. doi:10.1002/adma.200401260. http://doi.wiley.com/10.1002/adma.200401260.

Tseng, C. R., H. Y. Lee, and F. C. Chang. 2001. Crystallization kinetics and crystallization behavior of syndiotactic polystyrene/clay nanocomposites. *Journal of Polymer Science Part B: Polymer Physics* 39 (17) (September 1): 2097–2107. doi:10.1002/polb.1184. http://doi.wiley.com/10.1002/polb.1184.

Van Houten, D. J. and D. G. Baird. 2009. Generation of low-density high-performance poly(arylene ether sulfone) foams using a benign processing technique. *Polymer Engineering & Science* 49 (1) (January): 44–51. doi:10.1002/pen.21214. http://doi.wiley.com/10.1002/pen.21214.

Verdejo, R., P. Werner, J. Sandler, V. Altstädt, and M. S. P. Shaffer. 2009. Morphology and properties of injection-molded carbon-nanofibre poly(etheretherketone) foams. *Journal of Materials Science* 44 (6) (January 15): 1427–1434. doi:10.1007/s10853-008-3168-y. http://www.springerlink.com/index/10.1007/s10853-008-3168-y.

Vrentas, J. S. and J. L. Duda. 1977. Diffusion in polymer—Solvent systems. I. Reexamination of the free-volume theory. *Journal of Polymer Science: Polymer Physics Edition* 15 (3) (March): 403–416. doi:10.1002/pol.1977.180150302. http://doi.wiley.com/10.1002/pol.1977.180150302.

Wang, D., H. Gao, W. Jiang, and Z. Jiang. 2007a. Microcellular processing and relaxation of poly(ether ether ketone). *Journal of Polymer Science Part B: Polymer Physics* 45 (20) (October 15): 2890–2898. doi:10.1002/polb.21266. http://doi.wiley.com/10.1002/polb.21266.

Wang, D., H. Gao, W. Jiang, and Z. Jiang. 2007b. Effect of supercritical carbon dioxide on the crystallization behavior of poly(ether ether ketone). *Journal of Polymer Science Part B: Polymer Physics* 45 (21) (November 1): 2927–2936. doi:10.1002/polb.21150. http://doi.wiley.com/10.1002/polb.21150.

Wang, C., C. L. Huang, Y. C. Chen, G. L. Hwang, and S. J. Tsai. 2008. Carbon nanocapsules-reinforced syndiotactic polystyrene nanocomposites: Crystallization and morphological features. *Polymer* 49 (25) (November): 5564–5574. doi:10.1016/j.polymer.2008.09.057. http://linkinghub.elsevier.com/retrieve/pii/S0032386108008409.

Wang, D, W. Jiang, H. Gao, and Z. Jiang. 2006. Diffusion and swelling of carbon dioxide in amorphous poly(ether ether ketone)s. *Journal of Membrane Science* 281 (1–2) (September 15): 203–210. doi:10.1016/j.memsci.2006.03.047. http://linkinghub.elsevier.com/retrieve/pii/S0376738806002201.

Wang, X., W. Tong, W. Li, H. Huang, J. Yang, and G. Li. 2006. Preparation and properties of nanocomposite of poly(phenylene sulfide)/calcium carbonate. *Polymer Bulletin* 57 (6) (July 27): 953–962. doi:10.1007/s00289-006-0652-x. http://www.springerlink.com/index/10.1007/s00289-006-0652-x.

Weiser, E. S., T. F. Johnson, T. L. St. Clair, Y. Echigo, H. Kaneshiro, and B. W. Grimsley. 2000. Polyimide foams for aerospace vehicles. *High Performance Polymers* 12 (1): 1–12. doi:10.1088/0954-0083/12/1/301.

Wong, J. C. H., E. Tervoort, S. Busato, U. T. Gonzenbach, A. R. Studart, P. Ermanni, and L. J. Gauckler. 2010. Designing macroporous polymers from particle-stabilized foams. *Journal of Materials Chemistry* 20 (27): 5628. doi:10.1039/c0jm00655f. http://xlink.rsc.org/?DOI = c0jm00655f.

Wu, T. M., S. F. Hsu, C. F. Chien, and J. Y. Wu. 2004. Isothermal and nonisothermal crystallization kinetics of syndiotactic polystyrene/clay nanocomposites. *Polymer Engineering and Science* 44 (12) (December): 2288–2297. doi:10.1002/pen.20256. http://doi.wiley.com/10.1002/pen.20256.

Wu, S. Y., Y. L. Huang, C. C. M. Ma, S. M. Yuen, C. C. Teng, and S. Y. Yang. 2011. Mechanical, thermal and electrical properties of aluminum nitride/polyetherimide composites. *Composites Part A: Applied Science and Manufacturing* 42 (11) (November): 1573–1583. doi:10.1016/j.compositesa.2011.06.009. http://linkinghub.elsevier.com/retrieve/pii/S1359835X1100193X.

Wu, T. M. and C. Y. Liu. 2005. Poly(ethylene 2,6-naphthalate)/layered silicate nanocomposites: Fabrication, crystallization behavior and properties. *Polymer* 46 (15) (July): 5621–5629. doi:10.1016/j.polymer.2005.04.071. http://linkinghub.elsevier.com/retrieve/pii/S0032386105005495.

Wu, X., X. Liu, J. Wei, J. Ma, F. Deng, and S. Wei. 2012. Nano-TiO2/PEEK bioactive composite as a bone substitute material: *In vitro* and *in vivo* studies. *International Journal of Nanomedicine* 7 (January): 1215–1225. doi:10.2147/IJN.S28101. http://www.pubmedcentral.nih.gov/articlerender.fcgi?artid = 3298387&tool = pmcentrez&rendertype = abstract.

Xanthos, M., C. Wan, R. Dhavalikar, G. P. Karayannidis, and D. N. Bikiaris. 2004. Identification of rheological and structural characteristics of foamable poly(ethylene terephthalate) by reactive extrusion. *Polymer International* 53 (8) (August 29): 1161–1168. doi:10.1002/pi.1526. http://doi.wiley.com/10.1002/pi.1526.

Xanthos, M., M. W. Young, G. P. Karayanndis, and D. N. Bikiaris. 2001. Reactive modification of polyethylene terephthalate with polyepoxides. *Polymer Engineering & Science* 41 (4) (April): 643–655. doi:10.1002/pen.10760. http://doi.wiley.com/10.1002/pen.10760.

Xian, G., Z. Zhang, and K. Friedrich. 2006. Tribological properties of micro- and nanoparticles-filled poly(etherimide) composites. *Journal of Applied Polymer Science* 101 (3) (August 5): 1678–1686. doi:10.1002/app.22578. http://doi.wiley.com/10.1002/app.22578.

Yesil, S. and G. Bayram. 2011. Poly(ethylene terephthalate)/carbon nanotube composites prepared with chemically treated carbon nanotubes. *Polymer Engineering & Science* 51 (7) (July 11): 1286–1300. doi:10.1002/pen.21938. http://doi.wiley.com/10.1002/pen.21938.

Yudin, V. E. and V. M. Svetlichnyi. 2010. Effect of the structure and shape of filler nanoparticles on the physical properties of polyimide composites. *Russian Journal of General Chemistry* 80 (10) (November 26): 2157–2169. doi:10.1134/S1070363210100452. http://www.springerlink.com/index/10.1134/S1070363210100452.

Yun, S., Y. Heo, H. Im, and J. Kim. 2012. Sulfonated multiwalled carbon nanotube/sulfonated poly(ether sulfone) composite membrane with low methanol permeability for direct methanol fuel cells. *Journal of Applied Polymer Science* 126 (S2) (November 25): E513–E521. doi:10.1002/app.36741. http://doi.wiley.com/10.1002/app.36741.

Zhai, W., W. Feng, J. Ling, and W. Zheng. 2012. Fabrication of lightweight microcellular polyimide foams with three-dimensional shape by CO2 foaming and compression molding. *Industrial & Engineering Chemistry Research* 51 (39) (October 3): 12827–12834. doi:10.1021/ie3017658. http://pubs.acs.org/doi/abs/10.1021/ie3017658.

Zhang, G., L. Chang, and A. K. Schlarb. 2009. The roles of nano-SiO2 particles on the tribological behavior of short carbon fiber reinforced PEEK. *Composites Science and Technology* 69 (7-8) (June): 1029–1035. doi:10.1016/j.compscitech.2009.01.023. http://linkinghub.elsevier.com/retrieve/pii/S0266353809000207.

Zhang, J., Y. Zhang, Y. Chen, L. Du, B. Zhang, H. Zhang, J. Liu, and K. Wang. 2012. Preparation and characterization of novel polyethersulfone hybrid ultrafiltration membranes bending with modified halloysite nanotubes loaded with silver nanoparticles. *Industrial & Engineering Chemistry Research* 51 (7) (February 22): 3081–3090. doi:10.1021/ie202473u. http://pubs.acs.org/doi/abs/10.1021/ie202473u.

Zhang, H. B., W. G. Zheng, Q. Yan, J. W. Wang, Z. H. Lu, G. Y. Ji, and Z. Z. Yu. 2010. Electrically conductive polyethylene terephthalate/graphene nanocomposites prepared by melt compounding. *Polymer* 51 (5) (March 2): 1191–1196. doi:10.1016/j.polymer.2010.01.027. http://linkinghub.elsevier.com/retrieve/pii/S003238611000056X.

# Index

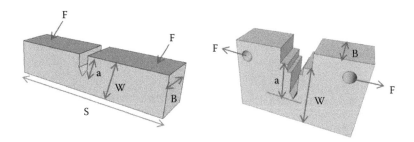

**FIGURE 2.2** Typical specimen geometries for SENB and CT tests.

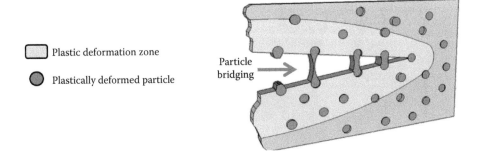

Plastic deformation zone

Plastically deformed particle

Particle bridging

**FIGURE 2.3** Schematic diagram of particle bridging-toughening mechanism. The bridging particles tend to close the crack while surrounding particles in the plastic deformation zone absorb fracture energy through plastic deformation.

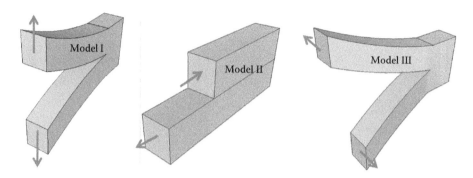

**FIGURE 2.5** Schematic representation for the crack modes.

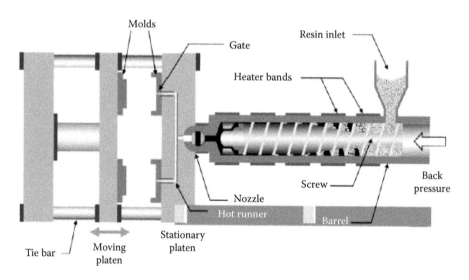

**FIGURE 5.2** Injection foaming process. (Reprinted from Lim L. T. et al., 2008, *Progress in Polymer Science* 33, 820–852. Copyright 2013, with permission from Elsevier.)

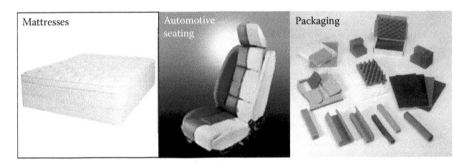

Mattresses

Automotive seating

Packaging

**FIGURE 6.4** Some of the possible applications of flexible PU foams.

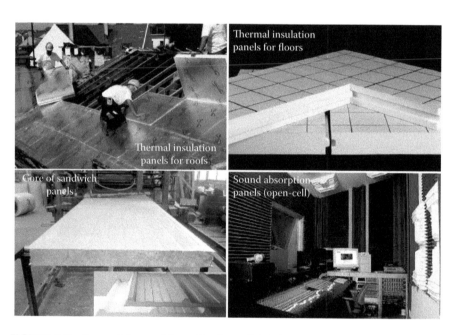

**FIGURE 6.5** Some of the possible applications of rigid PU foams.

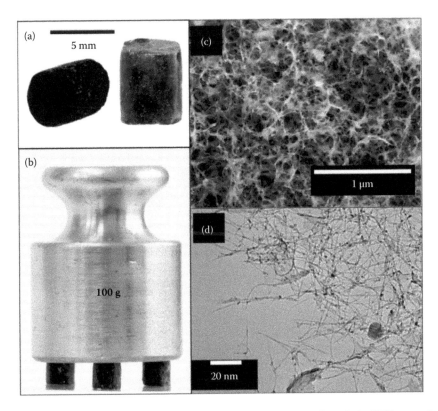

**FIGURE 8.1** Images of aerogels. (a) Macroscopic pieces of 7.5 mg/mL CNT aerogels. Pristine CNT aerogel (left) appears black, whereas the aerogel reinforced in a 1 mass% PVA bath (right) is slightly gray. (b) Three PVA-reinforced aerogel pillars (total mass = 13.0 mg) supporting 100 g, or ca. 8000 times their weight. (c) This scanning electron microscopy (SEM) image of a critical-point-dried aerogel reinforced in a 0.5 mass% PVA solution (CNT content = 10 mg/mL) reveals an open, porous structure. (d) This high-magnification transmission electron microscopy (TEM) image of an unreinforced aerogel reveals small-diameter CNTs arranged in a classic filamentous network. (Reprinted from Bryning M. B. et al., 2007, *Adv. Mater.* 19, 661–664, with permission from John Wiley & (Reprinted from Bryning M. B. et al., 2007, *Adv. Mater.* 19, 661–664, with permission from John Wiley & Sons.)

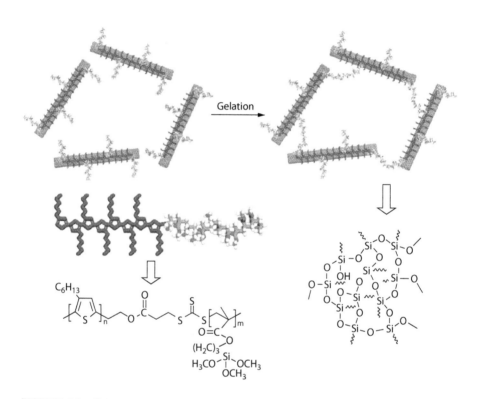

**FIGURE 8.3** Schematic illustration of gelation process of P3HT-b-PTMSPMA dispersed MWCNTs. Before gelation, P3HT blocks bond to the MWCNT surface through π–π interaction; PTMSPMA blocks locate at the outer surface of MWCNT. After gelation, MWCNTs interact with each other through chemical bonding formed by PTMSPMA blocks. (Reprinted from Zou J. et al., 2010, *ACS Nano* 4, 7293–7302, with permission from the American Chemical Society.)

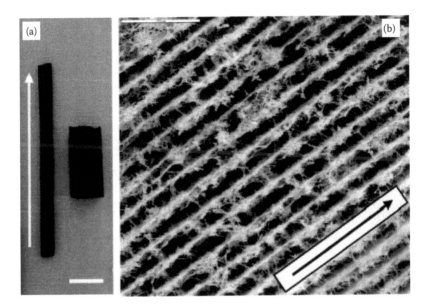

**FIGURE 8.5** Picture of MWCNT/CHI monoliths with different shapes and sizes resulting from the ISISA processing of MWCNT/CHI suspensions placed in different disposable containers; an insulin syringe (left) and a polystyrene cuvette (right). (a) Bar is 1 cm. SEM micrograph of the longitudinal section of a MWCNT/CHI monolith. (b) Bar is 50 μm. The MWNT content of every monolith shown is 85 mass%. Arrows indicate the direction of freezing. (Reprinted from Gutiérrez M.C. et al., 2007, *J. Phys. Chem. C* 111, 5557–5560, with permission from the American Chemical Society.)

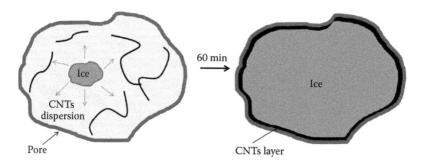

**FIGURE 8.7** Schematic representation of ice-mediated deposition of CNTs onto inner surfaces of cryogel walls. (Reprinted from Petrov P. and Georgiev G., 2011, *Chem. Commun.* 47, 5768–5770, with permission of The Royal Society of Chemistry.)